T0199353

# Modern Distribution Systems with PSCAD Analysis

# Modern Distribution Systems with PSCAD Analysis

Atousa Yazdani

CRC Press
Taylor & Francis Group
Boca Raton  London  New York

CRC Press is an imprint of the
Taylor & Francis Group, an **informa** business

CRC Press
Taylor & Francis Group
6000 Broken Sound Parkway NW, Suite 300
Boca Raton, FL 33487-2742

© 2018 by Taylor & Francis Group, LLC
CRC Press is an imprint of Taylor & Francis Group, an Informa business

No claim to original U.S. Government works

Printed on acid free paper

International Standard Book Number-13: 978-1-138-03355-9 (Hardback)

**Visit the Taylor & Francis Web site at**
**http://www.taylorandfrancis.com**

**and the CRC Press Web site at**
**http://www.crcpress.com**

*To the memory of Grandma, Florin Banapour*

# Contents

# *Preface*

A distribution system—as part of the power system—plays a crucial role in supplying everyday electricity demands. Operational functionality of distribution grids is imperative in meeting energy demands and maintaining customer satisfaction. Throughout the years, experts have published books designed for the purpose of assisting engineers and power system students in obtaining knowledge related to this system.

This book is designed to familiarize undergraduate students with distribution systems. Traditional theories and fundamental concepts required to analyze distribution systems are considered. After providing the classical basis of distribution system fundamentals, this text will incorporate information that is relative to the changing scope of the power distribution industry. The introduction of green energy generation, such as solar and wind, has revolutionized the topology, operation, and protection of the distribution system. The traditionally passive distribution grids are now being integrated with distributed generation and are functioning bidirectionally in terms of energy demand and supply. This book will include sections to facilitate understanding of these new advancements in distribution systems, familiarizing power system students with possible impacts associated with active distribution systems.

Lastly and most importantly, this book will offer power engineering students a chance to even further familiarize themselves with each of the specified concepts covered within the text by providing experimental modeling and analysis opportunities. This will be done by utilizing the commonly used software, PSCAD. Each topic will be analyzed and tested by this modeling and analysis tool, which will provide an abundance of examples to enhance the understanding of the modern distribution system.

MATLAB® is a registered trademark of The MathWorks, Inc. For product information, please contact:

The MathWorks, Inc.
3 Apple Hill Drive
Natick, MA 01760-2098 USA
Tel: 508 647 7000
Fax: 508-647-7001
E-mail: info@mathworks.com
Web: www.mathworks.com

# *Author*

I started my career at the age of 22 after receiving my undergraduate degree from Tehran University. They had hired me as a design engineer working on cabling and electrical distribution of high-voltage substations. While working as a design engineer managed to get my master's degree in Tehran Polytechnics. A couple of years later, I decided to continue my education in Missouri, United States, where I received my Ph.D. I have been involved in a number of consulting projects related to distribution systems, transients in power systems and renewable integration and their impacts on the system. I chose the career path of becoming a full-time faculty after spending 15 years working with engineers and consultants. This book is authored to help students get a grasp of the concepts and phenomena in distribution systems while experimenting them using PSCAD software.

# 1

## Introduction to Distribution System

## 1.1 Distribution System Introduction and Design Considerations

An electric power system, as it is shown in Figure 1.1, comprises a group of components including generation equipment, a transmission system, subtransmission, and distribution system. The major difference, when comparing the mentioned components, exists in their functionality. There are voltage differences when connecting the mentioned systems together, therefore electric power transformers are used to interconnect the power grid.

Distribution system design is done by answering questions related to

- Maximum capacity of the system that will be obtained by knowing the capacity and nameplate values on each device the amount of load that they can accommodate, can be designed.
- Operating limits (voltage limits and line capacity limits).
- Given the physical distance between generation sites to loads, the question will be: Is the system able to maintain voltage standards? Length of the installed lines and transformers nameplate capacity are useful to find the level of load they can accommodate.
- Increasing efficiency is a major concern to design engineers. The question here is how can power loss in the system be decreased to acceptable levels?
- Power quality enhancement and its incorporation in system design will answer the question if each customer is receiving constant levels of voltage and current. An example of poor power quality would be a customer who has a problem with flickering lights.
- Measurements and control: How will the system be monitored and controlled remotely?

And recently the integration of renewable energy in distribution grids has brought up new challenges and design questions for distribution system engineers.

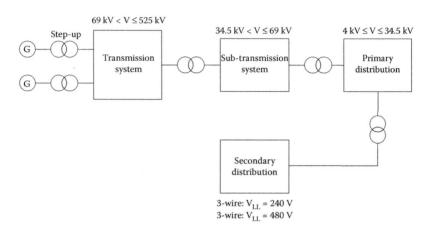

**FIGURE 1.1**
Electric utility system components and typical voltage levels.

Modeling with computer-aided simulation (among other design tools) is a technological development that significantly aids the design and maintenance process. After a system is designed and implemented, it is necessary to continuously collect data to verify that all systems are operating at scheduled levels, and to control mechanisms in the system for objectives such as voltage and current correction. Systems such as SCADA were designed with these purposes in mind. The most accurate systems collect data points using synchronous measurement, which measures both voltage and current phasors in real time, and presents the data as a time-stamped vectors.

In this book, we introduce several design and operating problems associated with distribution systems by utilizing simple models in PSCAD software.

Figure 1.2 is a simple sketch of a step-down substation. The required equipment for the substation comprises of protection schemes and control implementation. Step-down transformer with the implemented on load tap changer (LTC). Later on in this chapter, we will talk more about the voltage regulation and primary and secondary distribution system. The protection and control of the distribution system are not topics that will be discussed in this book but rather we will talk about operational concepts on the distribution system.

Figure 1.3 is showing a wood pole structure with three single-phase transformers installed on the top of the pole. The pole is actually a medium between the primary feeder and secondary feeder that is connected through three single-phase transformer banks. This application is for overhead lines. Figure 1.4 shows a pad-mount transformer used for supplying underground distribution lines [12,13].

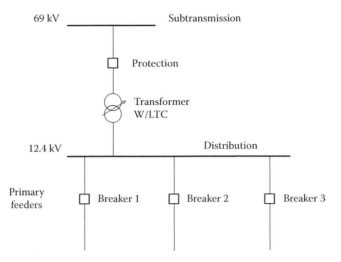

**FIGURE 1.2**
Simple substation configuration.

**FIGURE 1.3**
Overhead service transformer. (From 2017. "Distribution Transformer." https://en.wikipedia.org/wiki/Distribution_transformer.)

**FIGURE 1.4**
Pad-mount transformer (underground service transformer). (From 2017. "Padmount Transformer." https://en.wikipedia.org/wiki/Padmount_transformer.)

## 1.2 Simple Radial System

As it is shown in Figure 1.2, a simple radial system receives power at the source voltage at a single substation and the step-down transformer makes usable at the utilization level. This is called the primary distribution system. The voltage levels more than 4–15 kV are in this category. There are customers that are supplied directly from the primary distribution system. Depending on the size of the customer, they may own their required substations. But for houses and other small power applications another alternative exists, which is the secondary unit substation. In this case, the switchgear or switchboard is designed and installed as a close-coupled single assembly. Utilities are usually the owner of the primary equipment and transformer. They supply the customer at the utilization voltage, which is 120 V single-phase or 240 V double-phase.

Panelboards that are located closer to loads are connected to low voltage feeder circuits through switches and breakers. Due to the diversity of the load that is connected to the main transformer, it is possible to optimize the size of the transformer. This means that the loads that are connected to the feeders and eventually the transformer don't experience their maximum all at the same time.

In order to enhance the operation of distribution systems there are other design possibilities such as:

- *Loop primary system and radial secondary system*: This system has more than one "primary loop" and transformers connected. Interlock logics are built into the system to prevent closing all the switches in one loop and form a short-circuit path. Figure 1.5 shows this structure.

- *Primary selective system and secondary radial system*: In this design there are at least two primary feeders available. In case of a power loss in one of the main transformers, the switching can transfer the load to a healthy feeder. Figure 1.6 shows this structure.

- *Two-source primary and secondary selective system*: This is a similar design to the previous one. If there is a fault in one substation, the load will get transferred to the healthy substation. Also, normally there is an open tie at the secondary to provide redundancy of supply. Figure 1.7 shows this structure.

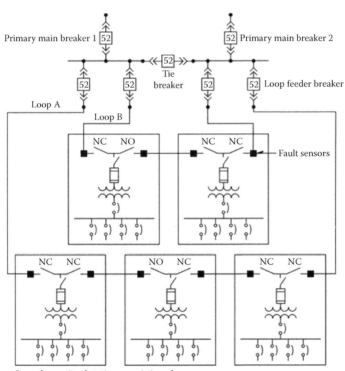

Secondary unit substations consisting of:
duplex primary switches/fused primary switches/transformer and secondary main feeder breakers

**FIGURE 1.5**
Loop primary system and radial secondary system. (From Eaton Consultants. August 2017. "Power Distribution Systems." No. CA08104001E, www.eaton.com/consultants.)

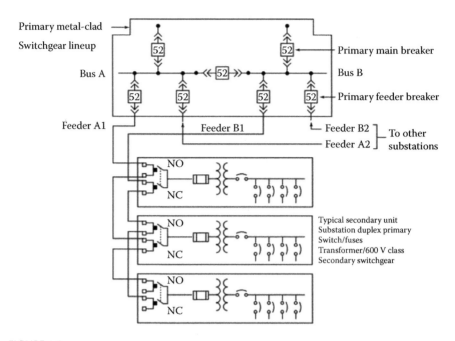

**FIGURE 1.6**
Primary selective system and secondary radial system. (From Eaton Consultants. August 2017. "Power Distribution Systems." No. CA08104001E, www.eaton.com/consultants.)

## 1.2.1 Essential Equipment in Distribution Systems

Following is a list of equipment that are essential for the sound operation of a distribution system.

- Poles
- Overhead transformers
- Switches and sectionalizes
- Circuit reclosers
- Fuses
- Capacitors
- Voltage regulators

The poles are responsible for holding the equipment above the ground. Transformers are responsible for changing the voltage levels from the primary side to the secondary side. A sectionalizer is designed to isolate faults on circuits in conjunction with reclosers. A sectionalizer does not interrupt fault current. It waits until the recloser de-energizes the line. After a predetermined number of operations by the recloser, the sectionalizer opens

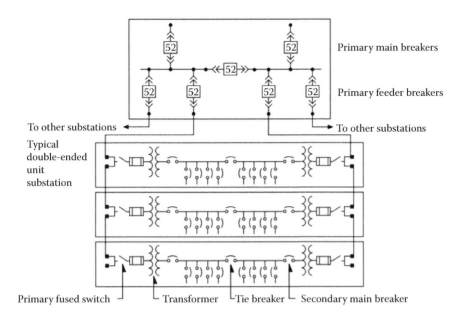

**FIGURE 1.7**
Two-source primary and secondary selective system. (From Eaton Consultants. August 2017. "Power Distribution Systems." No. CA08104001E, www.eaton.com/consultants.)

to isolate the section of faulty line. This allows the recloser to re-energize the line up to the open sectionalizer [4].

Figure 1.2 shows a path to transfer energy from high voltage to low voltage customers. A distribution line is a line or system for distributing power from a transmission system to a consumer that operates at less than 69 kV. This system contains different sizes of conductors based on the amount of energy that is expected to be transferred. Another important equipment in the system is switches and circuit breakers. With certain associated logics applied to circuit breakers, the distribution feeders will be protected. A distribution system is divided into two different voltage levels, called primary distribution and secondary distribution systems. A distribution system starts right after the step-down transformer with voltage levels less than 46 kV. The primary distribution system is a set of three-phase feeders that will get connected to service transformers as shown in Figure 1.8. Primary distribution substations can include LTCs, capacitor banks, and reclosers. The mentioned substations will provide load to three-phase and single-phase feeders.

Subtransmission lines with intermediate voltage levels are responsible for transferring power from the bulk power system to the distribution system. The medium voltage level power lines can get connected directly to large customers such as industrial and commercial consumers. The primary feeders are responsible for carrying energy from higher voltage to low voltage customer sites. The secondary system can be designed with three

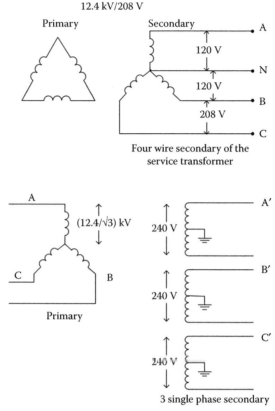

**FIGURE 1.8**
Three- and four-wired distribution system.

wires utilizing three-phase transformers and connecting their secondary to single-phase loads. This configuration is called a three-wire secondary configuration. It is possible to supply single-phase loads using single phase and neutral wires of distribution transformers. Three-phase, 380 V, four-wire secondary cable carries the electrical power from the distribution transformer to the building main switchboard. Figure 1.8 shows three-wire and four-wire distribution configurations [13].

It is not always necessary to include a subtransmission line—many utilities connect distribution substations directly to transmission lines.

Figure 1.9 shows a neighborhood of eight houses being supplied from a three-wire secondary system. Each has a double-phase feed and neutral feed.

Loads in a distribution system are either commercial or residential. The commercial loads are connected to the primary network, and residential

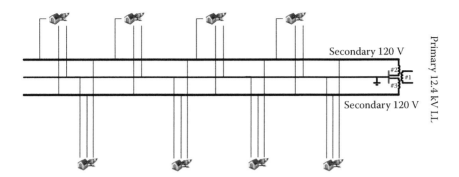

**FIGURE 1.9**
Three-wire secondary distribution system.

loads are supplied from the secondary. The industrial loads are usually motor loads, which are mostly induction machines. Residential loads consist of heating, cooling, and lightning and some other small motor loads. Household loads are mostly resistive loads with small reactive power requirement but industrial loads, on the other hand, demand a large amount of reactive power.

### 1.2.2 Line Models in Distribution Systems

Depending on the length of the lines different models have to be used for power flow analysis. In cases where the system voltage is less than 69 kV or the length of the line is less than 50 miles, the line is considered short and the capacitances of the lines will be ignored. Therefore, the impedance of the line with the length $l$ can be calculated as following:

$$Z = (r + j\omega L)l \ (\Omega),$$

where $r$ and $L$ are per phase resistance and reactance of the line.

The line charging will be concerning when the length of the line is more than 50 miles. In case the length $l$ lies in 50 miles $< l < 150$ miles, the line is considered medium length. In this case, a $\pi$ model is an accurate model for steady state and transient analysis. In Figure 1.10, a $\pi$ section is represented.

$$Y = (g + j\omega C)l \ (\Omega) \tag{1.1}$$

In this book, for the studies that we are conducting, the $\pi$ section model is used due to the short lengths of the distribution lines that are studied.

### 1.2.3 Load Modeling

For long distribution lines that are heavily loaded, an aggregate load contains a large number of individual devices. These devices are switching on and off

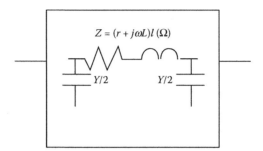

**FIGURE 1.10**
π section line.

many times during the day. For residential loads, the higher percentage at night is lighting, while for the annual peaks, the majority of the load is related to hot days (or depending on the region, very cold winter days).

The behavior of load changes with a change in the electrical parameters of the feeder (i.e., voltage and frequency) [14].

One of the common ways to model loads is ZIP coefficient load modeling. This model represents the active and reactive part of the load as a function of voltage. Typically, load surveys are performed on typical residential, commercial, and industrial customers to find the proportions of load in different categories as in ZIP In other words a load survey answers the following questions:

1. What percentage of the load is constant impedance (Z)?
2. What percentage of the load is constant current (I)?
3. What percentage of the load is constant power (P)?

The polynomial expression, known as the ZIP coefficients from Reference 5, represents the variation of load with respect to voltage.

In PSCAD, the load is modeled as a function of voltage magnitude and frequency, where the load real and reactive power are considered separately using the expressions:

$$P = P_0 \left( \frac{V}{V_0} \right)^{NP} (1 + K_{PF} \cdot dF) \tag{1.2}$$

$$Q = Q_0 \left( \frac{V}{V_0} \right)^{NQ} (1 + K_{QF} \cdot dF) \tag{1.3}$$

where,
$P$: Load real power
$P_0$: Rated real power per phase

$Q$: Load reactive power
$Q_0$: Rated reactive power per phase
$V$: Operating voltage
$V_0$: Rated voltage
$NP$: $dP/dV$ Voltage index for real power
$K_{PF}$: $dP/dF$ Frequency index for real power
$NQ$: $dQ/dV$ Voltage index for reactive power
$K_{QF}$: $dQ/dF$ Frequency index for reactive power

Depending on what is mentioned above, the parameters explained above are conventionally chosen from values in Table 1.1.

Figure 1.11 shows a system that is built in PSCAD to show load dependency to voltage variation at substation.

Figure 1.12 is showing how the RMS voltage at the substation is changing with time.

The system is rated at 34.5 kV. First the load is divided into 30% constant impedance, 30% constant current, and 40% constant power as shown in Figure 1.11 (i.e., diverse load). Figure 1.13 shows the load measurement at the substation for this condition. Then for the sake of comparison, the total load is changed to

1. Total constant impedance
2. Total constant current
3. Total constant power

Figure 1.14 makes a comparison for the active power that is measured at the substation with voltage variation for the above three cases along with when the load is diverse.

As it can be seen from Figure 1.14, the case where the total load is constant impedance shows a very high variation with changes in voltage. If the total load is constant power the voltage variation will show a minimal effect on the demanded power.

**TABLE 1.1**

Load Model Coefficients

|  | NP | K_PF | NQ | K_QF |
|---|---|---|---|---|
| Constant impedance | 2 | 2 | 0 | −1 |
|  | 2 | 2 | 0 | 0 |
|  | 2 | 2 | 0 | 1 |
| Constant current | 1 | 1 | 0 | 0 |
| Constant power | 0 | 0 | 0 | 0 |

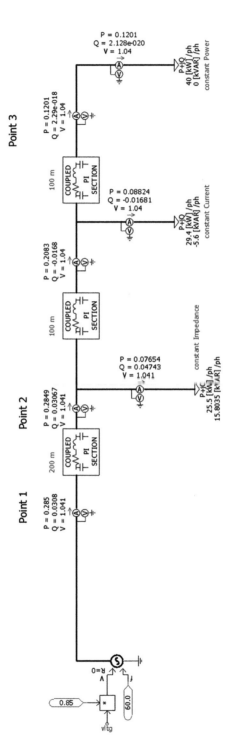

**FIGURE 1.11**
Load variation due to voltage variation.

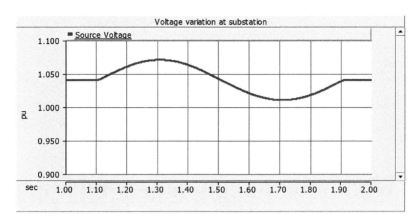

**FIGURE 1.12**
Applied voltage variation at the substation.

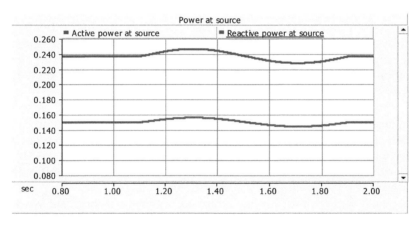

**FIGURE 1.13**
Active and reactive power for diverse loads.

## 1.3 Voltage Regulation

The process maintaining the voltage along a distribution feeder within the bandwidth specified by American National Standards Institute (ANSI), also known as ANSI-C84.1, is called voltage regulation. The standard for the low and high limits is as following

- Under normal operating conditions, the regulation requirement is ±5% on a 120-volt base.
- Under unusual conditions, the allowable range is −8.3% to +5.8%.

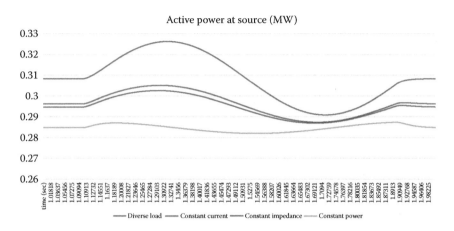

**FIGURE 1.14**

Comparison of different load models and the impact on the load supply at substation.

In a distribution feeder without active generation the highest voltage is at the substation voltage regulator (LTC). The lowest voltage is at the far away points from the substation. Both the low and high voltages in normal operation are kept within 114–126 V respectively (0.95–1.05 pu on a 120 V base) as stated by the ANSI Service Voltage Standard C84.1.

The Institute of Electrical and Electronics Engineers (IEEE) standard mandates the voltage along the distribution feeder to be more than 0.95 pu and less than 1.05 pu. In the case of long feeders with a high amount of load, the feeder can experience voltages less than the minimum indicated by the standard. Voltage regulators, which are autotransformers with multiple taps, are used to regulate voltage at places with low voltage.

A voltage regulator is simply a tapped autotransformer that works to either raise (boost) or lower (buck) the voltage. This device can be considered the most important device in the single line diagram in order to transfer energy to a long distance. It has the responsibility of connecting two points with different voltage levels. Each transformer has the following specified nameplate values: kilo Volt Ampere (kVA) rating, voltage rating, impedance, no-load loss, and saturation condition. Transformers in a distribution system can be three-phase connected as a Y or delta. Also, they can be formed as three separate single-phase transformers.

In a distribution system, the main transformer usually contains an LTC. As it is shown in Figure 1.15, the secondary of the transformer with tap changer has multiple taps.

Here, to expand on the idea of voltage regulators, we start with a simple single-phase autotransformer. An autotransformer is a transformer with single coil as shown in Figure 1.16.

An autotransformer consists of a single winding which serves as primary winding as well as secondary winding. The winding AB of total turns $N_1$

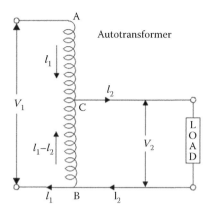

**FIGURE 1.15**
Single-phase autotransformer.

is considered as primary winding. This winding is tapped from point "C" and the portion BC is considered as secondary. The secondary of the autotransformer has $N_2$ turns. Therefore, the part AC has $N_1$–$N_2$ turns.

Figure 1.16 shows a PSCAD representation of a single-phase autotransformer. As it can be seen the active power is oscillating due to the single-phase load connection. The average of the active power delivered to the load is a lot higher. Figure 1.17 shows another simulation in this regard.

The multimeter on the secondary of the transformer is measuring active power, reactive power, and voltage at load. In the case of an autotransformer the output power is 5.5 MW but with the single-phase transformer the output is only 6 kW.

Figure 1.18 shows a single-phase transformer used as an autotransformer. Essentially by connecting a point on the secondary of the single-phase transformer to a point on the primary of the transformer, the autotransformer is formed. The device is utilizing a single coil to change voltage and transfer energy. As it is shown above, the single-phase transformer can supply a load of 24.96 kVA but the autotransformer is supplying a 773.76 kVA load.

A key advantage to the autotransformer is higher output power. The disadvantage is lower impedance, which means in higher voltage applications it may have a high short-circuit current.

**EXAMPLE 1.1**

A single-phase transformer, as follows, is going to be used as an autotransformer to supply a load. Here we are going to calculate some electrical quantities associated with the autotransformer knowing the amount of energy that is going to be supplied to the load.

$$I_1 = \frac{1000}{34.5} = 28.98 \text{ A}$$

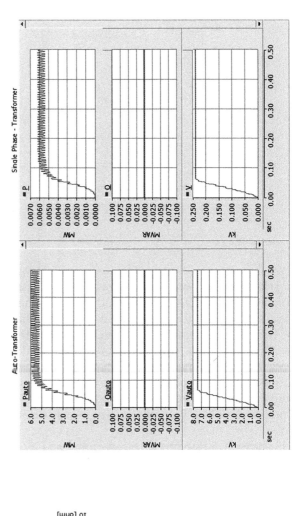

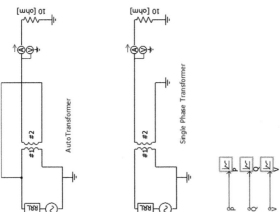

**FIGURE 1.16**
Single-phase transformer versus autotransformer.

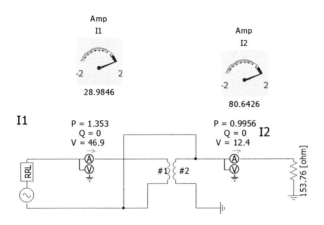

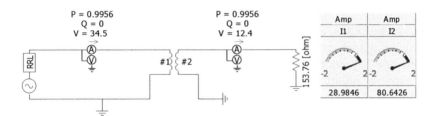

**FIGURE 1.17**
Autotransformer 2.

$$I_2 = \frac{1000}{12.4} = 80.64 \text{ A}$$

$$I_c = I_1 - I_2 = -51.66 \text{ A}$$

$$\text{Transformation Ratio: } \frac{34.5}{12.4} = 2.875$$

$$S_1 = (34.5 - 12.4) \times 28.98 = 640 \text{ kVA}$$

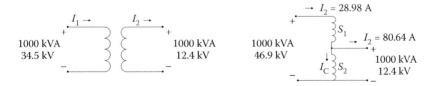

**FIGURE 1.18**
Single-phase transformer and autotransformer.

$$S_2 = 51.66 \times 12.4 = 640 \text{ kVA}$$

$$\Rightarrow S_{eq} = 640 \text{ kVA}$$

$$\text{Transferred kVA} = (V_1 I_2 - S_{eq}) = 34.5 \times 28.98 - 640 = 359.81 \text{ kVA}$$

An autotransformer has higher efficiency than a two-winding transformer. This is because of less ohmic loss and less core loss due to reduction of transformer material.

An autotransformer has better voltage regulation as the voltage drops in resistance and reactance of the single winding is less.

Autotransformers are often used when voltage transformation is small and cost is a considerable concern. Despite the advantages of voltage regulation, larger output power, elimination of secondary windings, and power transformation energy transfer, common drawbacks of using an autotransformer include risks of high short-circuit currents (due to a much smaller impedance) and complications due to the primary windings not being completely insulated from the secondary.

The special autotransformer, called the voltage regulator, uses the same principle of creating a direct electrical connection between primary and secondary windings, except that it also maintains the capability of changing the transformer turns ratio to meet changing voltage demands. It is designed so that the connection between the primary and secondary can change locations. This is done by switching the high potential connection on the secondary side between different connection points, called "taps." Switching between different taps creates a range of turn ratios that can be utilized, thus allowing the voltage at the secondary to be controllable and variable. For example, referring to Figure 1.19, the selector could connect with the tap associated with $N_6$ and the overall turns ratio would become $N_1:N_6$. If, for example, $N_1:N_6 = 1$ pu, then raising or lowering the tap selector to taps above or below the tap associated with tap $N_6$ could raise or lower the voltage $V_s$ in respect to $V_p$.

Voltage regulators are devices that maintain distribution voltage within a specified range of values. They are used by power companies to minimize

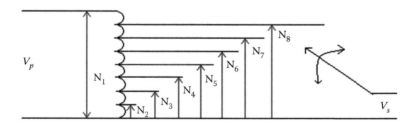

**FIGURE 1.19**
Example voltage regulator.

voltage drop and to ensure that the proper voltage is supplied to customers. Voltage regulators work in the range of ±10% of voltage increase or decrease. This range is usually divided into 16 steps or taps. Each tap can increase or decrease the voltage 5.8%.

**EXAMPLE 1.2**

A 69 kV/13.8 kV transformer is used as a voltage regulator. The voltage at the primary is measured 5% less than the rated value. If the secondary tap is on 1.03 what is the voltage at the secondary of the transformer?

$$V_{sec} = \frac{0.95 \times 69 \times 13.8 \times 1.03}{69} = 13.503 \text{ kV}$$

What is the normal secondary current and how does it change with the tap, when the transformer is rated at 500 kVA?

At the rated primary voltage and without the tap, the current on the secondary will be:

$$I_{normal} = \frac{500}{\sqrt{3} \times 13.8} = 20.9 \text{ A}$$

With the tap and less voltage on the primary, the current at the secondary will be:

$$I_{sec} = \frac{500}{\sqrt{3} \times 13.5} = 21.383 \text{ A}$$

## 1.3.1 Voltage Regulators and Voltage Drop Calculation

Figure 1.20 shows a simple feeder to transfer energy to a load. The load is characterized as a passive load with the combination of a resistive and an inductive part. The line is identified with R (ohms) and X (ohm). The assumption is that the line is short enough to ignore shunt capacitors.

KVL is written for the feeder and it is shown in the following equations. Since the load is inductive the phasor representation of the current delivered to the load is written as the phasor $(\tilde{I}_R - j\tilde{I}_X)$; where $j = \sqrt{-1}$

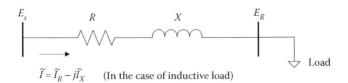

**FIGURE 1.20**
Simple feeder for voltage drop calculation.

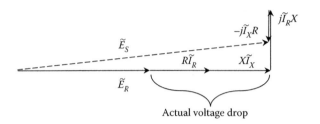

**FIGURE 1.21**
Phasor diagram for voltage drop calculation.

$$\tilde{E}_S = \tilde{E}_R + \tilde{I}(R + jX) \tag{1.4}$$

$$\tilde{E}_S = \tilde{E}_R + (\tilde{I}_R - j\tilde{I}_X)(R + jX) \tag{1.5}$$

$$\tilde{E}_S = \tilde{E}_R + R\tilde{I}_R + X\tilde{I}_X - j\tilde{I}_X R + j\tilde{I}_R X \tag{1.6}$$

Figure 1.21 is showing the real part of the voltage drop calculation and the reactive part is neglected. Therefore, voltage drop formula can be simplified as in 4.

$$\text{Voltage Drop (Line} - \text{Neutral)} = \tilde{E}_S - \tilde{E}_R = R\tilde{I}_R + X\tilde{I}_X = I(R\cos\theta + X\sin\theta) \tag{1.7}$$

It is important to pay attention to the fact that the voltage drop calculation is performed in a per-phase situation.

### EXAMPLE 1.3

If voltage drop is defined as $V_D = I(R \cos\theta + X \sin\theta)$, what happens to the voltage when a 7.5-mile piece of conductor with a loop impedance of $0.060 + j0.090$ $\Omega$/mile supplies load changed from 150 A real to 150 A inductive?

$$R = 0.060 \times 7.5 = 0.45 \ \Omega$$

$$X = 0.090 \times 7.5 = 0.675 \ \Omega$$

a. 150 A resistive load $\cos(\theta) = 1$ $V_D = 150(0.45 \times 1 + 0.675 \times 0) =$ 67.5 V
b. 150 A reactive load $\cos(\theta) = 0$ $V_D = 150(0.45 \times 0 + 0.675 \times 1) =$ 101.25 V

If the system is Y connected, and the base voltage is 12 kV how much in percentage is the voltage drop?

$$V_D = \frac{67.5}{12000/\sqrt{3}} = 0.974\%$$

$$V_D = \frac{101.25}{12000/\sqrt{3}} = 1.461\%$$

What is the voltage drop effect if the base of the voltage is 4.0 kV?

$$V_D = \frac{67.5}{4000/\sqrt{3}} = 2.922\%$$

$$V_D = \frac{101.25}{4000/\sqrt{3}} = 4.384\%$$

**EXAMPLE 1.4**

The voltage at the secondary side of the transformer is 12.4 kV LL. Calculate the voltage drop quantity for the feeder shown in Figure 1.22 considering the voltage drop formula as follow:

$$V_D = I(R\cos\theta + X\sin\theta)$$

Find:

a. Total voltage drop
b. What is the total active power at the substation
c. Total reactive power at the substation
d. Find S (complex power) at the substation
e. Power factor at the substation

The following MATLAB code is written to calculate the system voltage drop.

```
clear all
clc

R1=0.02;
R2=0.01;
R3=0.035;
```

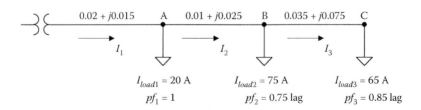

**FIGURE 1.22**
Distribution feeder with three loads.

```
X1=0.015;
X2=0.025;
X3=0.075;

j=sqrt(-1);
Iload1= 20;
Iload2= 75*0.75-j*75*sin(acos(0.75));
Iload3= 65*0.85-j*65*sin(acos(0.85));

I1=Iload1+Iload2+Iload3; %current after Tr to point A
I2=Iload2+Iload3;         %current for section AB
I3=Iload3;                %current for section BC

S=sqrt(3)*12.4*conj(I1) %apparent power at the secondary of
   the transformer
P=real(S) %active power at the secondary of the transformer
Q=imag(S) %reactive power at the secondary of the
   transformer

cosPhi=P/abs(S) %power factor at substation

theta1=abs(angle(I1));
theta2=abs(angle(I2));
theta3=abs(angle(I3));

I1=abs(I1);
I2=abs(I2);
I3=abs(I3);

Vdrop1=I1*((R1)*cos(theta1)+(X1)*sin(theta1)); % voltage
% drop between TR and A
Vdrop2=I2*((R2)*cos(theta2)+(X2)*sin(theta2));%voltage drop
   between AB
Vdrop3=I3*((R3)*cos(theta3)+(X3)*sin(theta3)); %voltage
   drop between BC

Vdrop=Vdrop1+Vdrop2+Vdrop3
```

Running the code copied above, the voltage drop for each section of the line is calculated as following:

```
Vdrop1 = 3.8877V
Vdrop2 = 3.2112V
Vdrop3 = 4.5018V
```

The voltage drop of the whole system is the addition of the all the calculated voltage drops:

```
Vdrop_total=11.6008 V
S = 2.8243 +j 1.8009 kVA
P = 2.8243 kW
Q = 1.8009 kVAR
cosPhi = 0.8432 lag
```

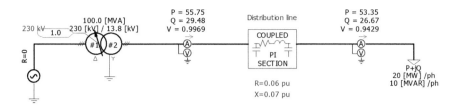

**FIGURE 1.23**
Voltage drop calculation built in PSCAD.

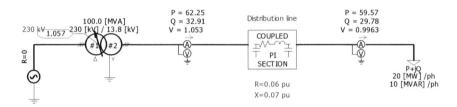

**FIGURE 1.24**
Voltage compensation at load node by increasing the transformer tap.

### EXAMPLE 1.5

Figure 1.23 shows a system that is built in PSCAD. Calculate the feeder voltage drop at the load. Compensate the voltage drop at the load by adjusting the voltage regulator tap on the transformer.

The data shown on the single line diagram of Figure 1.23 are the active power, reactive power, and per-unit voltage for each node. As it can be seen from the diagram, voltage at the load is 0.9429 pu. This shows a $1 - 0.9429 = 0.057$ pu voltage drop in order to push the voltage at the load to 1 pu, a possible solution is to use the transformer as an LTC and change the tap ratio on the secondary to 1.057. Figure 1.24 shows the results of the voltage fix at the load by increasing the tap 5.7%.

### EXAMPLE 1.6

Consider the following distribution feeder shown in Figure 1.25. Power factors for all the loads are equal to the value of 0.75 lag. Voltage at the secondary of the transformer is 480 V. The impedance of the line AB is j0.0125 Ω/mile and length of the line is 50 mile.

1. Find the total load that the source needs to supply.
2. Find the approximate voltage drop associated with the line AB.

The following MATLAB code solves the above-mentioned problem.

```
clear all;
clc;

j=sqrt(-1);
```

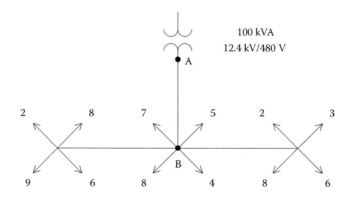

**FIGURE 1.25**
Distribution feeder example.

```
S= [2 8 6 9 7 5 8 4 2 3 8 6];
pf= 0.75*ones(1,12);
length=50;
z=j*0.0125*length;

sum=0;
theta1 = acos(0.75)

for i=1:12
   sum=sum+S(i);
end

x=abs(z);
r=0;
I=sum/(sqrt(3)*0.480);
VD_AB=I*(r*cos(theta1)+x*sin(theta1))
```

The voltage drop at AB will be 33.8124 V (phase voltage drop) which comparing the value of 33.8124 V to 480 V (line-to-line) voltage at the secondary of the transformer, there is 12.12% voltage drop. The reason why the voltage drop is so high is the fact that the reactive power of the loads is high due to small power factors. Try to solve the problem considering different power factors for each load.

**EXAMPLE 1.7**

Consider the following distribution feeder that has been built and ran in PSCAD. The lines are considered to be pi models with very small shunt capacitors (neglecting the suseptance of the lines). The system is connected to an ideal 12.4 kV source through an ideal 100 kVA, 12.4/4 kV Δ Y transformer. Lines are considered to be in per-unit length so the impedances shown here are the total impedance of each piece of line.

The multimeters measure the active and reactive power in MW and MVAR, respectively. The voltages are measured in per-unit.

We are trying to answer the following questions about the circuit.

a. Find voltage drops for each piece of the line
b. How much capacitor bank is needed to be installed at the substation to compensate all the reactive power that is supplied by the source?
c. If the substation transformer has a tap changer, find a steady state tap that fixes the voltage at the third load to 0.98 pu.

For part "a" it is possible to calculate the voltage drops in per-unit seeing the results from PSCAD model as follow:

$$Vd1 = 1 - 0.9781 = 0.0219 \text{ pu} = 0.0219 \times 4000/\sqrt{3} = 50.57 \text{ V}$$
$$Vd2 = 0.9781 - 0.9713 = 0.0068 \text{ pu} = 0.0068 \times 4000/\sqrt{3} = 15.7039 \text{ V}$$
$$Vd3 = 0.9713 - 0.9668 = 0.0045 \text{ pu} = 0.0045 \times 4000/\sqrt{3} = 10.3923 \text{ V}$$

For part "b" we change the PSCAD graph to Figure 1.26 with an attached Y connected capacitor bank sized 0.1753 MVAR. This will compensate almost all the reactive power which is needed to be supplied by the source. Some of them are related to the loads and some are the reactive power loss on the lines.

For part "c" essentially we are trying to push the voltage 0.8825–0.95 pu.

$$Tap = 1 + (0.98 - 0.9668) = 1.0132$$

Running the case again without the cap bank and with the voltage regulator tap calculated earlier yields the results shown in Figure 1.27.

### 1.3.2 Voltage Profile

Voltage drop is always a concern about distribution feeders. One of the ways to see how the voltage decreases from the substation by increasing the length and the load is to plot the voltage profile of the feeder. Here we use the example above and expand the idea of voltage profile.

For the above example, we try to plot the voltage profile of the feeder. If the lines are 5 miles, 3 miles and 2 miles from the substation to the last load the voltage profile of the feeder will be as drawn in Figure 1.28.

Figure 1.29 shows an example feeder by IEEE. It is called IEEE 34 bus test system. The feeder is accommodated with two voltage regulators.

Figure 1.30 shows the feeder voltage profile. The blue line is showing a case where no voltage regulator has reacted yet. The voltage at the end of the feeder can go as low as 0.9313 pu the orage figure shows the feeder voltage profile.

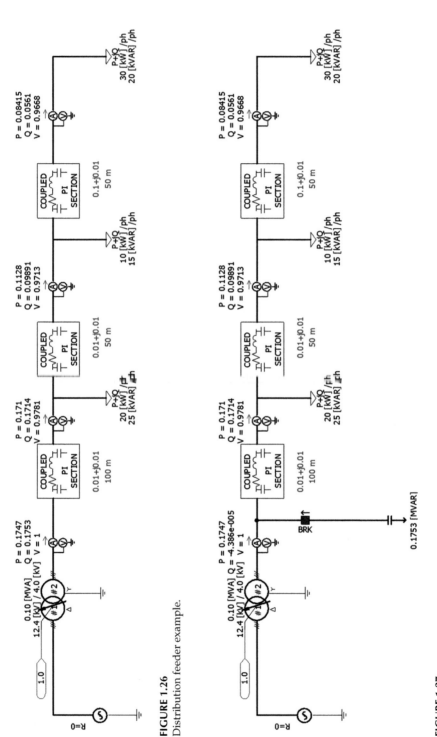

**FIGURE 1.26**
Distribution feeder example.

**FIGURE 1.27**
Reactive power compensation at the substation.

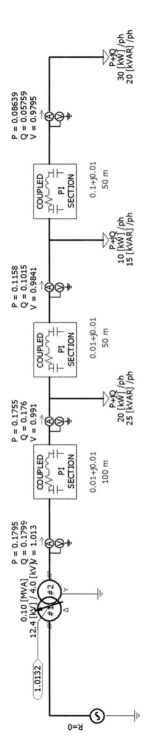

**FIGURE 1.28**

Voltage regulation at the substation.

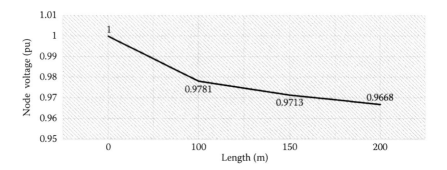

**FIGURE 1.29**
Feeder voltage profile.

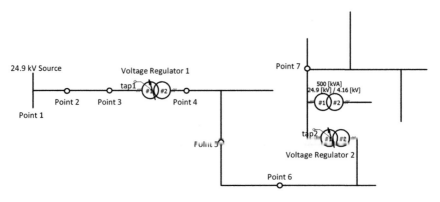

**FIGURE 1.30**
Simplified IEEE 34 bus test feeder.

## 1.4 Capacitor Banks

Capacitor banks are the providers of reactive power and they are used to enhance and compensate node voltages along the distribution feeders. There are several ways to control the switching of capacitor banks to the distribution feeders.

1. Automatic by sensing the voltage at the installation node and closing or opening the cap bank.
2. Time control, in this case the cap bank will be switched at a certain time. This is scheduled due to the history of the feeder and perhaps the time of maximum load.
3. The switching of capacitor banks can be done manually.

Capacitor banks are installed single-phase or three-phase. The nameplate values associated to them is kVAR and the rated voltage.

Many factors in a distribution system result in voltage levels that either rise too far above or too far below scheduled voltage levels. Transmitting current over long distances results in voltage drops due to losses over lines. Voltage sag may also be caused by switching operations associated with a temporary disconnection of supply, the flow of inrush currents associated with the starting of motor loads, or the flow of fault currents. Open circuit (open load), decrease in reactive power absorption at a load, and lightning, among other causes, can create voltage rises in a distribution system. In order to compensate for these voltage sags and rises to reasonable levels as quickly and effectively as possible, voltage regulation techniques and procedures must be implemented.

One common tool used for voltage level correction is capacitor banks. During normal operation, the level of the reactive power (Q [VAR]) in a grid system is the main controller of voltage level. This is because voltage levels are directly proportional to the reactive power in a grid. For example, if the major loads in the system are inductive loads, the voltage will drop at the load demand points, which also slightly affects the voltage at other points in the grid. This is because inductive loads absorb Q [VAR]. Voltage regulating capacitor banks would then be used to compensate this voltage drop (e.g., by injecting Q [VAR] into the grid).

Pole-mounted, and/or switched capacitors are responsible for power factor correction. Power factor correction along the feeder will enhance the feeder voltage profile. Depending on the expected daily load curves, groups of switched and fixed capacitors are installed along distribution feeders. The places which are accommodating of very highly inductive loads are the best place to install power factor correcting capacitors. A rule of thumb in a distribution system is to install the capacitor bank roughly two-thirds of the distance down the substation on feeders with a consistent load profile.

Consider the example that we did for the voltage regulation in Figure 1.26, add a capacitor bank of size 0.2 MVAR to the point shown in Figure 1.31, and draw the feeder voltage profile with and without the capacitor bank (Figures 1.32 and 1.33).

### 1.4.1 Power Factor Correction with Capacitors

Power factor is defined as the ratio of the real power to the apparent power at the same node. In other words, this is the cosine of the angle between current and the voltage at the same node. The more reactive power is needed by the system load, the smaller the power factor will be. Capacitor banks are installed to compensate for lines reactive power and push power factor to unity.

In an electric power system, a load with a low power factor draws more current than a load with a high power factor for the same amount of useful power transferred. The higher currents increase the energy lost in the

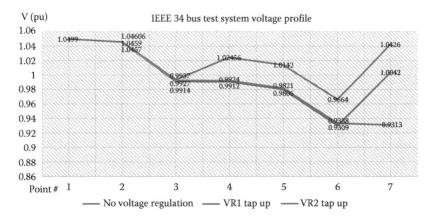

**FIGURE 1.31**
Voltage profile without and with voltage regulation.

distribution system, and require larger wires and other equipment. Because of the costs of larger equipment and wasted energy, electrical utilities will usually charge a higher cost to industrial or commercial customers where there is a low power factor. Factories have low power factors due to the installation of many inductive loads.

**EXAMPLE 1.8**

The power factor ($pF$) in a 100 kVA load is lowered from 0.9 to 0.7 (which is 22%). How much does active and reactive power at the load change? (Find $\Delta P$ and $\Delta Q$)

$$\theta_{old} = \cos^{-1}(pF_{old}) = \cos^{-1}(0.9) \approx 25.84°$$
$$\text{and} \quad \theta_{new} = \cos^{-1}(pF_{new}) = \cos^{-1}(0.7) \approx 45.57°$$

This shows that as the $pF$ decreases, $\theta$ increases. So, this scenario's power triangle transformation should look like the following (Figure 1.34):

*Note*: Notice since the kVA is constant in this case $P$ decreases and $Q$ increases as $\theta$ increases and pF decreases.

Initially:

$$P_{old} = |S|\cos\theta_{old} = |S|\,pF_{old} = (100\,\text{k})(0.9) = 90\,[\text{kW}]$$

$$Q_{old} = |S|\sin\theta_{old} = |S|\sqrt{1-(pF_{old})^2} = (100\,\text{k})\sqrt{1-(0.9)^2} \approx 43.58\,[\text{kVAR}]$$

After $pF$ change:

$$P_{new} = |S|\cos\theta_{new} = |S|\,pF_{new} = (100\,\text{k})(0.7) = 70\,[\text{kW}]$$

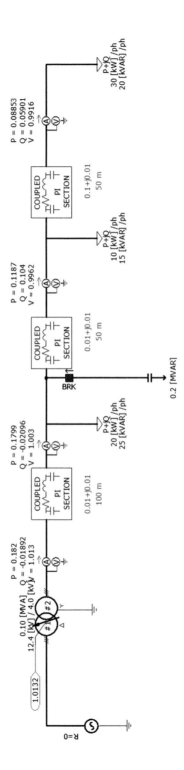

**FIGURE 1.32**

Voltage compensation with capacitor bank.

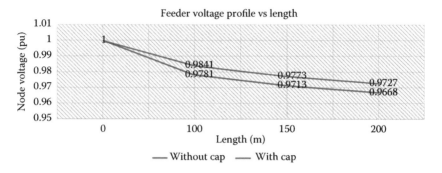

**FIGURE 1.33**
Voltage profile comparison with and without 0.2 MVAR capbank.

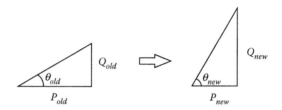

**FIGURE 1.34**
Power triangle.

$$Q_{new} = |S|\sin\theta_{new} = |S|\sqrt{1-(pF_{new})^2} = (100\,k)\sqrt{1-(0.7)^2} \approx 71.414\,[\text{kVAR}].$$

Therefore:

$$\Delta P = |P_{old} - P_{new}| = 90\,k - 70\,k = 20\,[\text{kW}]$$

and

$$\Delta Q = |Q_{old} - Q_{new}| \approx |71.414\,k - 43.58\,k| \approx 27.83\,[\text{kVAR}]$$

**EXAMPLE 1.9**

In a 250 kVA system, the power factor is changed from 0.82 to 0.95 but real power remains constant. What is the change in reactive power and apparent power? (Find $\Delta Q$ and $\Delta|S|$). Old (Figure 1.35):
   Let's use the following MATLAB code to solve the problem.

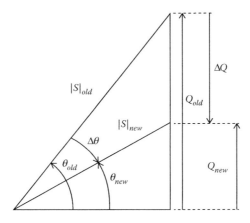

**FIGURE 1.35**
Power triangle for power factor correction.

```
clear all
clc;

pf_old=0.82;
pf_new=0.95;

kVA=250;

sinTheta_old=sqrt(1-pf_old^2);
sinTheta_new=sqrt(1-pf_new^2);

Qold=kVA*sinTheta_old;
Qnew=kVA*sinTheta_new;

Dq=Qnew-Qold
```

The change in reactive power will be calculated to be 75.71 kVAR.

**EXAMPLE 1.10**

A system is working at 250 kVA with a power factor of 0.85 lag.

a. How much reactive power is needed to change the power factor to 0.92 lag? If the system apparent power is kept constant?
b. How much reactive power is needed to change the power factor from 0.82 lag to 0.95 lag?

```
clear all
clc;

pf_old=0.85;
pf_new=0.92;

kVA=250;

sinTheta_old=sqrt(1-pf_old^2);
sinTheta_new=sqrt(1-pf_new^2);
```

```
Qold=kVA*sinTheta_old;
Qnew=kVA*sinTheta_new;

Dq1=abs(Qnew-Qold)

pf_old=0.82;
pf_new=0.95;

kVA=250;

sinTheta_old=sqrt(1-pf_old^2);
sinTheta_new=sqrt(1-pf_new^2);

Qold=kVA*sinTheta_old;
Qnew=kVA*sinTheta_new;
Dq2=abs(Qnew-Qold)
```
$\Delta$Q1 = 33.7161 kVAR
$\Delta$Q2 = 65.0284 kVAR

This example highlights the fact that as the power factor approaches 1, more and more reactive power must be injected to get the power factor closer to unity.

Traditionally, distribution feeders are long since the generation plants are far away from load centers. Also, the utilized voltage levels for distribution systems are very low. Voltages around 4 kV are still common in urban areas. With the above-mentioned conditions, it is required to take advantage of voltage regulators and/or capacitor banks to sustain voltages close to 1 pu throughout distribution feeders.

## 1.5 Distribution System Conductors

Load density and the physical conditions of the demand area will determine if the distribution lines are going to be overhead or underground. [https://ieeelongisland.org/pdf/viewgraphs/automating_power_distribution_system.pdf]

A very common form of quantifying the size of the conductors that are used in distribution systems is the standard American Wire Gauge (AWG).

AWG, also known as the Brown & Sharpe wire gauge, is a standardized wire gauge system used since 1857 predominantly in North America for the diameters of round, solid, nonferrous, electrically conducting wire. Dimensions of the wires are given in ASTM standard B 258. The cross-sectional area of each gauge is an important factor for determining its current-carrying capacity.

Increasing gauge numbers denote decreasing wire diameters, which is similar to many other nonmetric gauging systems such as Standard Wire Gauge (SWG). This gauge system originated in the number of drawing operations used to produce a given gauge of wire. Very fine wire (e.g., 30 gauge)

**TABLE 1.2**

Standardized Wire Gauge Used Since 1857 in North America for Round Conducting Wires

| Size | D (mm) | R Ω/1000 ft | Ampacity (A) | Fusing Current for 10 sec. |
|------|--------|-------------|--------------|----------------------------|
| 4/0 | 11.684 | 0.04901 | 230 at 75°C | 3.2 kA |

required more passes through the drawing dies than 0 gauge wire did. Manufacturers of wire formerly had proprietary wire gauge systems; the development of standardized wire gauges rationalized selection of wire for a particular purpose.

The AWG tables are for a single, solid, round conductor. The AWG of a stranded wire is determined by the cross-sectional area of the equivalent solid conductor. Because there are also small gaps between the strands, a stranded wire will always have a slightly larger overall diameter than a solid wire with the same AWG.

AWG is also commonly used to specify body piercing jewelry size. AWG gauges are also used to describe stranded wire. In this case, it describes a wire which is equal in cross-sectional area to the total of all the cross-sectional areas of the individual strands; the gaps between strands are not counted. When made with circular strands, these gaps occupy about 10% of the wire area, thus requiring a wire about 5% thicker than equivalent solid wire.

Stranded wires are specified with three numbers, the overall AWG size, the number of strands, and the AWG size of a strand. The number of strands and the AWG of a strand are separated by a slash. For example, a 22 AWG 7/30 stranded wire is a 22 AWG wire made from seven strands of 30 AWG wire [6] (Table 1.2).

Larger than 4/0 wires are identified by thousands of circular mils (kcmil), where 1 kcmil = 0.5067 mm². After 4/0 → 250 kcmil.

## 1.6 Grounding Equipment, System Grounding, and Grounding Transformer

Grounding is an essential part of electrical distribution system design and construction, due to safety considerations and requirement for proper operation of the system and equipment supplied by it. System grounding connects the electrical supply, from the utility, from transformer secondary windings, or from a generator, to ground. A system can be solidly grounded (no intentional impedance to ground), impedance grounded (through a resistance or reactance), or ungrounded (with no intentional connection to ground).

System grounding affects the susceptibility of the system to voltage transients, also will dictated the types of loads the system can accommodate, and more importantly it will provide information as how to protect the system.

Grounding determines:

- The amount of system voltage transient
- The type of load that can be accommodated
- Required protection

National Electric Code (NEC)

- For the utility grounding—grounding is done on the secondary winding
- On generators—grounding is on the stator configuration
- Transformers—grounding is on the on the winding configuration
- Static Power Compensator

### Solidly Grounded

This occurs when the ground impedance is very small, approximately zero. Also, it is located at the most common location for each phase.

The system grounding arrangement is determined by the grounding of the power source.

Consider the following figure showing a set of three-phase balanced voltages. The neutral in the system is not grounded and, therefore in the case of a single phase to ground fault on phase, the voltages for the healthy phases can be as large as $\sqrt{3}$ times the phase voltage. This can be detrimental to human beings as well as equipment (Figure 1.36).

Figure 1.37 shows several possible forms of grounding. In industry, there are other types of grounding that may cause unbalanced cases in the system (Figure 1.38).

The type of grounding in the system will affect the amount of voltage rise that the system may experience in transients. The IEEE standard C62.92 defines the following coefficient to classify system grounding. The system is called effectively grounded when

$$0 < \frac{X_0}{X_1} < 3, 0 < \frac{R_0}{R_1} < 1 \tag{1.8}$$

### Reactance Grounding

This is normally used in the neutrals of generators. The reactor will limit the fault current at the generator terminals. If the reactor is sized properly, the system will be considered effectively grounded.

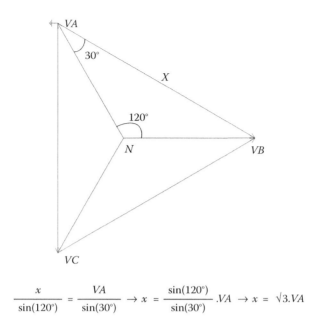

$$\frac{x}{\sin(120°)} = \frac{VA}{\sin(30°)} \rightarrow x = \frac{\sin(120°)}{\sin(30°)}.VA \rightarrow x = \sqrt{3}.VA$$

**FIGURE 1.36**
Single phase to ground fault ungrounded three-phase system.

The grounding conductor system is not intended to carry operational current. This path is intended to carry unwanted and fault currents for protection. The winding configurations will identify the type grounding that the system can have.

### Resistance Grounded

In general, where residually connected ground relays are used (51N), the fault current at each grounded source should not be limited to less than the current transformers rating of the source. This rule will provide sensitive differential protection for wye-connected generators and transformers against line-to-ground faults near the neutral. Of course, if the installation of ground fault differential protection is feasible, or ground sensor current transformers are used, sensitive differential relaying in resistance grounded system with greater fault limitation is feasible (Figure 1.39).

### 1.6.1 Grounding with Zigzag Transformer

#### Grounding Transformer to Mitigate the High Voltages on the Healthy Phases

The zigzag transformer will only pass ground current. Its typical implementation on an ungrounded system, in order to convert the system

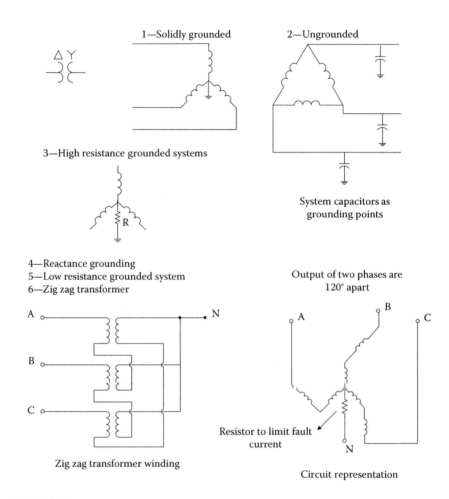

**FIGURE 1.37**
Types of system grounding.

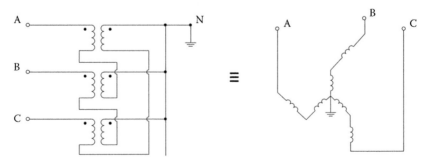

**FIGURE 1.38**
Zigzag transformer built from three single-phase transformers.

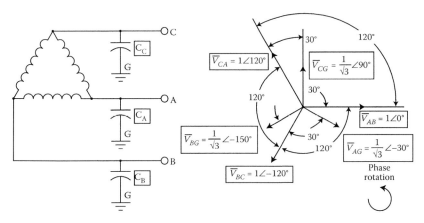

**FIGURE 1.39**
Ungrounded system with intrinsic capacitors as grounding points.

to a high-resistance grounded system, is shown in Figure 1.40. The zigzag transformer distributes the ground current equally between the three phases. For all practical purposes the system, from a grounding standpoint, behaves as a high-resistance grounded system. The following PSCAD example illustrates the fact that providing a ground path with zigzag transformer

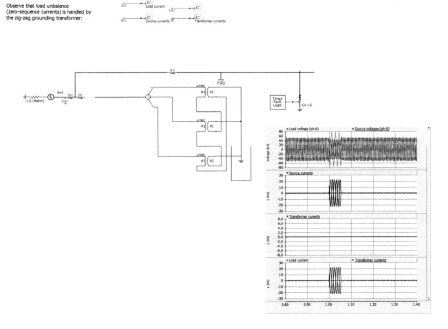

**FIGURE 1.40**
Double phase to ground fault without grounding transformer.

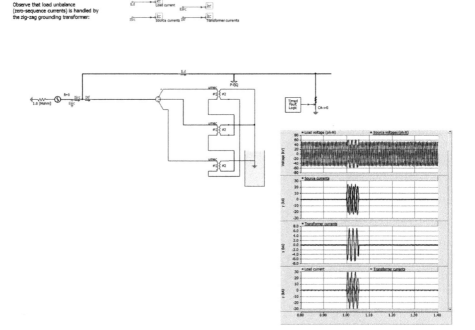

**FIGURE 1.41**
Double phase to ground fault with grounding transformer.

will protect the system by decreasing the transient overvoltages due to an unbalanced fault.

The zigzag transformer will only pass ground current. The zigzag transformer in the following example shows the ground path and the effect on the transient voltage on the healthy phase. In the absence of the grounding transformer the load voltage on phase B (healthy phase) reaches 80 kV while the normal peak is 50.5 kV. Installation of the grounding zigzag transformer limits the voltage rise to 60 kV (Figure 1.41). Zigzag transformer can be used for harmonic reduction purposes as well.

Figure 1.42 shows a case where grounding is added to the middle point of the delta connected winding. The phasor diagram shows the amount of voltage imbalance that can be caused by this type of grounding. The grounded point provides a path to ground in case of fault but will deteriorate the normal operation of the system due to unbalanced voltages. There are number of disadvantages associated with this grounding, such as when the system is not suitable for supplying single-phase loads. As mentioned, it will be very unsymmetrical in normal operation. Due to the increased voltages on healthy phase there is a risk for shock.

When there is a single phase to ground fault in the system, the fault current is low and may stay undetected. Therefore, the system may stay in operation

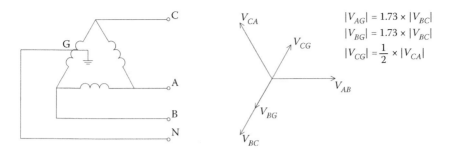

**FIGURE 1.42**
Arbitrary grounding on Δ connected transformer.

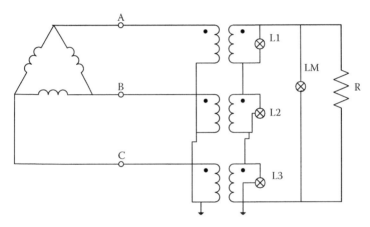

**FIGURE 1.43**
Ground fault detection method for ungrounded systems.

although there is a single-phase fault. However, it is required by the NEC standard that proper protections detect and isolate the fault.

The danger of the ungrounded system experiencing a single phase to ground fault is the increase in healthy voltages.

Ungrounded systems may experience Ferro resonance and that will cause high transient overvoltage. We will explain this phenomenon later in Chapter 2.

Single phase to ground fault can be detected utilizing the configuration introduced in Figure 1.43 and Table 1.3.

If the grounding at the neutral point is done with the addition of resistance or reactance, it can be categorized as follows with these associated characteristics.

- Low resistance
  - Limit fault current
  - Minimizes the damage at the fault

- High resistance
  - Operates during single phase to ground fault
  - Enables current to flow in protection path
  - Low current and therefore no damage
- Reactor connected
  - Reduces the ground fault current phase and ground
  - Higher fault current than low resistance
  - Better coordination in multi-grounded system

Figures 1.44 and 1.45 shows the grounding point on the secondary of the distribution transformer.

**TABLE 1.3**

Fault and Lights Condition

| Ground Fault Location | L1 | L2 | L3 | LM |
|---|---|---|---|---|
| None | Dim | Dim | Dim | Off |
| Phase A | Off | Bright | Bright | Bright |
| Phase B | Bright | Off | Bright | Bright |
| Phase C | Bright | Bright | Dim | Bright |

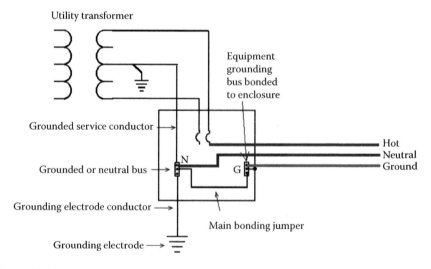

**FIGURE 1.44**
Grounding on the secondary of distribution transformer. (From Peele, S. "Grounding of Electrical Systems." Progress Energy, https://www.progress-energy.com/assets/www/docs/business/Grounding.pdf.)

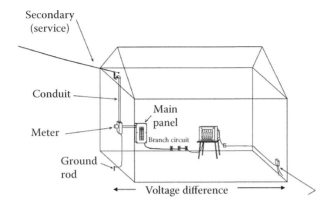

**FIGURE 1.45**
Grounding and bounding inside residential places. (From Peele, S. "Grounding of Electrical Systems." Progress Energy, https://www.progress-energy.com/assets/www/docs/business/Grounding.pdf.)

### *Grounding Electrode NEC 250.50*

Standards used for system grounding design are NEC and IEEE Green Book 142-1991 "Recommended Practice for Grounding of Industrial and Commercial Power Systems." And IEEE Emerald Book 1100–1999 "Recommended Practice for Powering and Grounding Sensitive Electronic Equipment."

### 1.6.2 Effective Grounding

In the case of solidly grounding a transformer, neutral large fault currents can flow through the transformer neutral. In order to mitigate this issue, the neutral of the transformer will be grounded through an impedance; this will limit the fault current while limiting the overvoltage on the healthy phase. The IEEE Effective Grounding as defined in IEEE Green Book is when

X0: zero sequence reactance

R0: zero sequence resistance

X1: positive sequence reactance

When the above conditions are met for conventional generators, the line-to-ground, short-circuit current will be limited to 60% of the three-phase, short-circuit value. Also, the overvoltage in the unfaulted phase will be limited to 140% of its nominal value, which is considered the limit of overvoltage without causing equipment damage.

Grounding can be done either by a zigzag or wye-delta transformer. The grounding transformer will provide a high impedance for positive sequence and low impedance for zero sequence currents. Extra impedance will be added to the transformer to make sure the above-mentioned relationships between X0 and X1 exist.

---

## 1.7 Loss Calculation

Power system losses happen in conductors, equipment such as transformers, distribution lines, and magnetic losses in transformers. The amount of demand in the network and also the electrical characteristics of the pieces mentioned earlier determine the amount of loss in the system.

Distribution lines on the primary and the secondary dissipate major amounts of energy.

The unexpected load increase was reflected in the increase of technical losses above the normal level.

There are two types of losses in the system. First, constant losses such as open circuit loss of transformers. Second is the variable loss, which is the active power loss in the network and can be calculated with the following formula.

$$\text{loss} = I^2 R \, (W)$$

As it can be seen, the loss is proportional to the resistive element of lines and also the amount of current squared. In this part, we have set up the following primary distribution feeder in PSCAD with the voltage of 12.4 kV. The feeder contains three pieces of line, each 1 mile in length, and the loop impedance of $0.01 + j*0.01$ $\Omega$/mile. Figure 1.46 shows the power flow on the system with the total load of $30 + j*30$ kVA/ph attached to the end of the line. Figure 1.47 repeats the case when the load is installed on the 2/3 of the feeder line. Figure 1.48 the load is attached to the 1/3 of the feeder. Figure 1.49 distributed the load, as shown, in every 1/3-length of the line. Table 1.4 summarized the loss associated with each case.

As can be seen in Table 1.5, the losses associated with Case 3 is less than other cases which makes sense because the load is closest to the supply. The best option after this is to distribute the load evenly across the feeder. This case also shows less losses comparing to Cases 1 and 2.

Here we repeat the cases mentioned above with adding cap banks 1/3 of the total load reactive power. Figures 1.50 through 1.53 repeat previous cases 1 through 4 with addition of capacitor bank with the size of 1/3 of load reactive power.

Table 1.5 summarizes the results illustrated on the figures. Again with the cap bank the optimal case if case 4 with loads and capacitor banks being

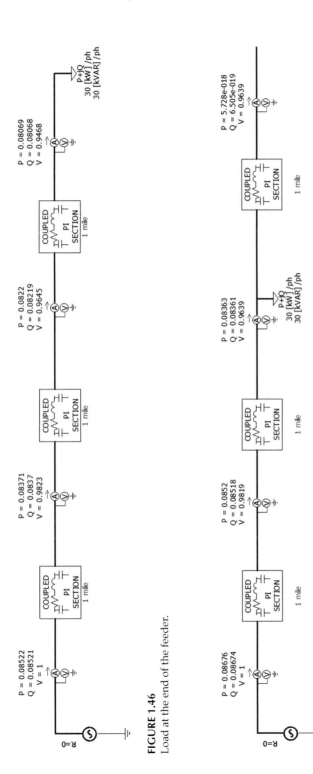

**FIGURE 1.46**
Load at the end of the feeder.

**FIGURE 1.47**
Load at 2/3 of the feeder.

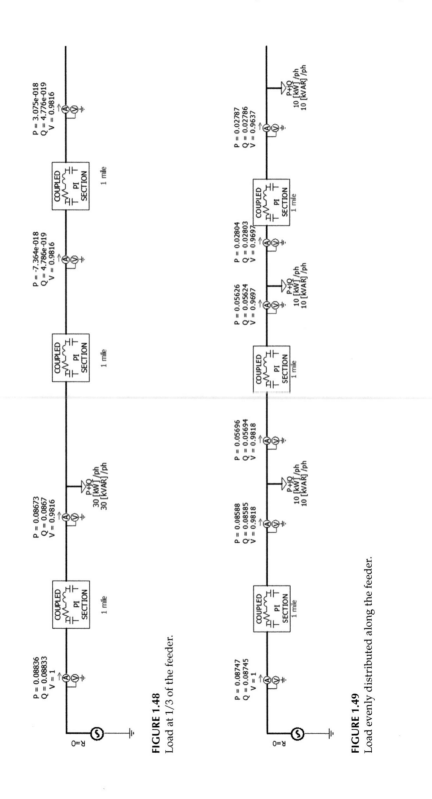

**FIGURE 1.48**
Load at 1/3 of the feeder.

**FIGURE 1.49**
Load evenly distributed along the feeder.

**TABLE 1.4**

Power Loss without Capacitor Banks

| Point of Measurement | Active Power (MW) | | | |
| --- | --- | --- | --- | --- |
| | Case 1 | Case 2 | Case 3 | Case 4 |
| 1 | 0.08522 | 0.08776 | 0.08836 | 0.08747 |
| 2 | 0.08371 | 0.0852 | 0.08673 | 0.08588 |
| 3 | 0.0822 | 0.08363 | 0 | 0.05626 |
| 4 | 0.08069 | 0 | 0 | 0.02787 |
| Ploss (MW) | 0.00453 | 0.00413 | 0.00163 | 0.00246 |

**TABLE 1.5**

Power Loss with Capacitor Banks

| Point of Measurement | Active Power (MW) with Cap Compensation | | | |
| --- | --- | --- | --- | --- |
| | Case 1 | Case 2 | Case 3 | Case 4 |
| 1 | 0.08547 | 0.08694 | 0.08845 | 0.08761 |
| 2 | 0.08436 | 0.08579 | 0.08727 | 0.08645 |
| 3 | 0.08325 | 0.08465 | 0 | 0.05684 |
| 4 | 0.08214 | 0 | 0 | 0.02821 |
| Ploss (MW) | 0.00333 | 0.00229 | 0.00118 | 0.0018 |

distributed along the feeder. Comparing the results of Case 4 in Tables 1.4 and 1.5. It can be seen that there is a major loss reduction with cap bank additions around 36%.

## Practice

Repeat the Case 4 and have only one capacitor bank with the size 1/3 of the total load at 1/3 of the line, 2/3 of the line and end of the line. Summarize your observations.

### EXAMPLE 1.11

A three-phase express feeder has an impedance of $8 + j18 \ \Omega/ph$. At the load end, the line-to-line voltage is 11.6 kV and the total three-phase power is 1500 kW at a lagging power factor of 0.78. The system is connected to a higher voltage of 34.5 kV through a transformer that has name plate values of 34.5/11.8 kV and 2 MVA. If the transformer impedance is 0.05 pu (Figure 1.54).

Find:

a. Line-to-line voltage at sending end
b. Power factor at sending end
c. Copper loss of the feeder
d. Power at sending end

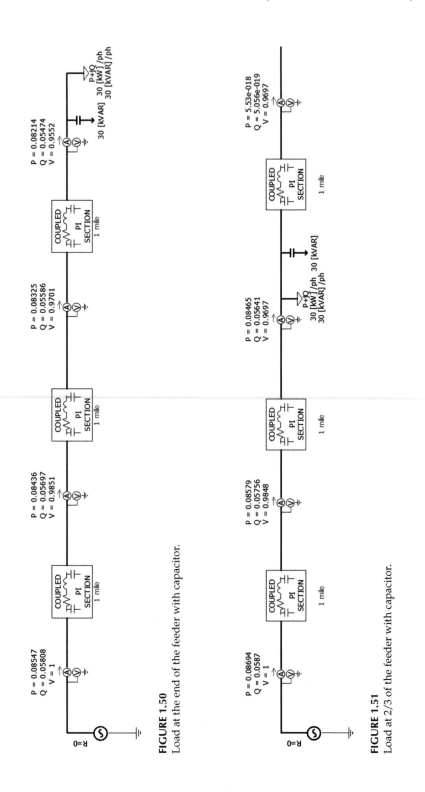

**FIGURE 1.50**
Load at the end of the feeder with capacitor.

**FIGURE 1.51**
Load at 2/3 of the feeder with capacitor.

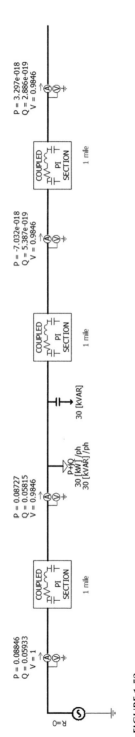

**FIGURE 1.52**
Load at 1/3 of the feeder with capacitor.

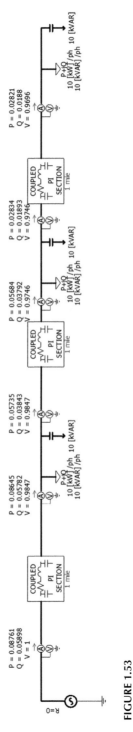

**FIGURE 1.53**
Load evenly distributed along the feeder with capacitor.

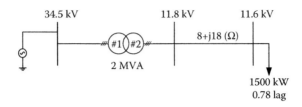

**FIGURE 1.54**
Single line diagram of example (ExamplePerUnitCalc.m.).

```
clear all;
clc;
j=sqrt(-1);
Vr=11.6; %kV

Pload=1.5; %MW
pfload=0.78;
Zline=8+j*18; %ohm
Ztr=0.05; %perunit

Vbase=11.8; %kV
Vr_pu=Vr/Vbase;
Sbase=2; %MVA
Zbase=Vbase^2/Sbase;
Ibase=Sbase/(sqrt(3)*Vbase);
Zline_pu=Zline/Zbase;

I=Pload/(sqrt(3)*Vr*pfload); %kA
Ipu=I/Ibase; %perunit
Iload=Ipu*pfload+j*Ipu*sqrt(1-pfload^2); %perunit

Ztotal=Zline_pu+Ztr;

Vs=Vr_pu+Ztotal*Iload; %perunit
Vs=Vs*Vbase; %kV

abs(Vs) %a) Line-to-line voltage at sending end in Ohm
cos(angle(Vs)-angle(Iload)) %b) Power factor at sending end
real(Zline)*(abs(Iload)*Ibase)^2%c) Copper loss of the feeder
    in MW
real(sqrt(3)*Vs*conj(Iload*Ibase)) %d) Power at sending end

Vsending = 11.7562 kV

pf = 0.9315

loss = 0.0733 MW

Pin =1.8155 MW
```

## PROBLEMS

1. What are the subsystems in the power system? How are these
   subsystems differentiated from each other?

2. A single-phase load is going to be supplied at the secondary of a distribution system. The load is a constant current load of 10 A with $\cos(\varphi) = 0.9$ lag. Find the active power absorbed by the load if the system supplies a three-wired or four-wired load?

3. What are the main components in a distribution system?

4. What is LTC and how does it work?

5. The following is showing a center-tapped transformer with the secondary of 480 V, the number of winding turns on the second secondary is 120. The first secondary has 7400 turns. What is the number of turns on the primary and what is the voltage across the second secondary coil? (Figure P1.1)

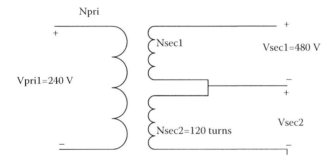

**FIGURE P1.1**
Autotransformer.

6. Use PSCAD. In the following circuit set the voltage at 1.05 and 0.95 pu. Use the load as in the following table (Figure P1.2)

|  | NP | K_PF | NQ | K_QF |
|---|---|---|---|---|
| Constant | 2 | 2 | 0 | −1 |
| impedance | 2 | 2 | 0 | 0 |
|  | 2 | 2 | 0 | 1 |
| Constant current | 1 | 1 | 0 | 0 |
| Constant power | 0 | 0 | 0 | 0 |

Therefore, you will have 10 cases. Measure the loss of the total graph and see which case is associated with the least amount of loss.

7. Use PSCAD. Build a distribution feeder with four major loads. The substation voltage is 13.8 kV. Assume the voltage is fixed at 1 pu at the substation. Distribute the load in a way that the voltage at the final load is 0.93 pu. Come up with possible solutions to increase the 0.93 pu to at least 0.97 pu. The solution cannot deteriorate the operational limits of the feeder. Draw the voltage profile of your feeder.

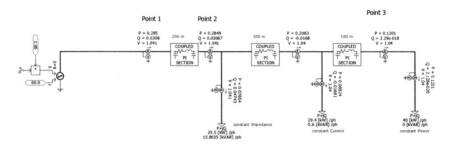

**FIGURE P1.2**
Network example.

8. Build the following figure in PSCAD. Apply single-phase faults at the primary. Do you think the lights will work to show the type of fault? (Figure P1.3)

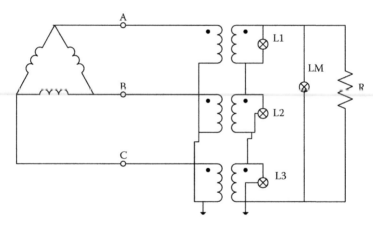

**FIGURE P1.3**
Ground fault identifier.

# 2

## Distribution System Analysis

## 2.1 Transients in Distribution System

The only way to analyze short duration phenomenon in any electric circuit is to solve the differential equations related to that system. The way PSCAD software is implemented is to solve a combination of differential equations, and, therefore, it is possible to see the effects of fast transients on the system electrical quantities. The question is: "Why do these fast transients, and analysis of their effects, play an important role?" The answer is in maintaining the personnel and equipment safety. Occurrence of a fast transient can induce a voltage traveling wave, with a magnitude many times higher than the system rated values. Although they may last in the order of microseconds, still the impulse voltages, better known as the transient overvoltages (TOVs), caused by the phenomenon are detrimental to personnel and equipment. Therefore, it is essential to understand the root causes of TOVs in designing a system and provide solutions to mitigate possible TOVs. Following is a list of incidents that cause TOVs in any distribution or transmission circuit.

- Lightning
- Capacitor switching
- Ferro resonance
- Line-to-line or unbalanced faults

Hereafter, we analyze each phenomenon and see the impact on an example PSCAD case.

### 2.1.1 Lightning

Depending on where our system stays geographically, it may be susceptible to be hit by several lightning surges during a year. In order to analyze the effect of lightning impulses, the surge is modeled as the curve illustrated in Figure 2.1.

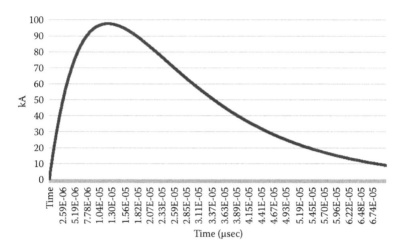

**FIGURE 2.1**
100 kA, 12 μsec × 36 μsec.

The waveform above shows an 8 × 20 current wave. It takes 12 μS to reach maximum, and 36 μS to decrease to half of the maximum. One important characteristic on the impulse waveform is the peak current value.

### Rate of Rise

If the maximum on the above 8 × 20 wave is at 100 kA, the rate of rise in that waveform will be 100 kA × 8 μS = 12.5 kA/1 μS.

### 2.1.1.1 Example of TOV Due to Lightning Impulse

Figure 2.2 shows a simple distribution system experiencing lightning. The objective is to insert the surge illustrated in Figure 2.3 into the system and analyze the effect of this fast transient on current and voltage at measurement Point 2. Figures 2.4 and 2.5 show the impact of the fast transient surge on voltage and current of the circuit. The voltage and current for unaffected phase (B and C) are the same. Therefore, they are plotted on top of each other. An important point here is the fact that the rated voltage in the system is 12.4 kV and with the surge the AC voltage on Phase A has reached to 32.5 MV. This is just an example of how dangerous the surge can be. This kind of voltage will be detrimental to all the equipment in the circuit. Later, we will try to mitigate the situation by adding surge arresters to the circuit.

### 2.1.2 Capacitor Switching

Capacitor banks are utilized in distribution feeders to enhance voltage profile along the feeder. Also in many applications, such as wind farms,

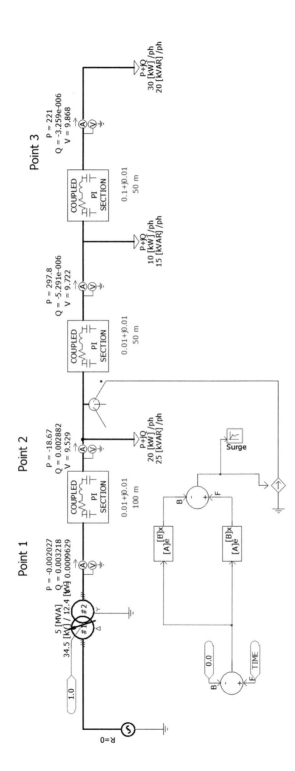

**FIGURE 2.2**
Distribution system with lightning impulse effect.

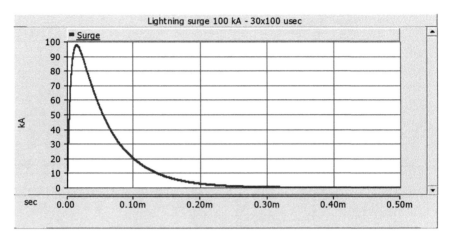

**FIGURE 2.3**
Surge function.

existence of inductive machines makes the utilization of reactive power support (i.e., shunt cap banks) inevitable. Here, we are trying to show the effect of switching a capacitor bank in the feeder as shown in Figure 2.6. The capacitor bank of 100 kVAR is being switched to point at $t = 0.2$ sec. Figures 2.7 and 2.8 show the three-phase AC voltages measured at Points 2 and 3. The rated voltage of the circuit is 4.0 kV. Therefore, the maximum voltage per phase will be $\left(4/\sqrt{3}\right) \times \sqrt{2} = 3.2659$ kV. Figure 2.6 shows the cap bank switching has pushed the maximum of the voltage to 5.45 kV, this means a 66.6% increase [13].

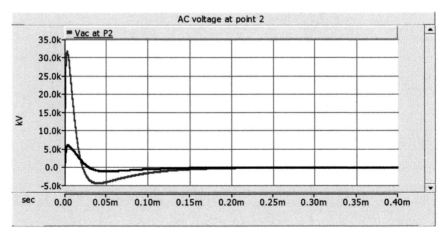

**FIGURE 2.4**
Three-phase voltages measured at Point 2.

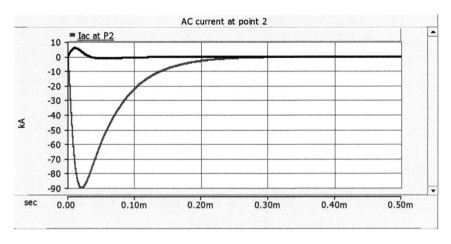

**FIGURE 2.5**
Three-phase currents measured at Point 2.

### 2.1.3 Ferro Resonance (FR)

Consider the combination of three-phase lines with intrinsic capacitor banks connected to a transformer as shown in Figure 2.9. The system is experiencing a fault on Phase A, and the line breaker operates only for the faulted line and keeps the other two-phase connected. If the transformer goes into saturation due to environmental effects or the fault current. There will be a case that the impedance of Phase A of the transformer will be the same as the impedance of the shunt cap bank on Phase A. By looking at Figure 2.9 again, when $Xc = Xtr$, it will be translated electrically on the circuit as if the Point A is solidly grounded. Our transformer in this case is Y-connected ungrounded. Now the circuit situation is changed to what is explained previously in Figure 2.10. Therefore, the healthy phases can experience voltage rises up to $\sqrt{3}$ of their normal condition. There are systems that are more susceptible to this such as in the case of small transformers, which saturate faster; primary system underground and shielding increases the amount of cap, and secondary cap with floating neutral.

There are obvious solutions for the phenomenon such as

- Installing wye-wye grounding at neutral and connected to the primary neutral. This way the series circuit between the cap and magnetizing inductance will be short circuited even if one of the phases becomes open.
- Installation of three-pole breakers and preventing single pole operations.
- Not to switch the line and transformer together and keep the m separate.

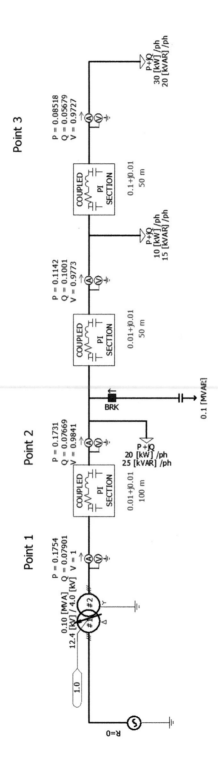

**FIGURE 2.6**
Distribution system with capacitor bank switching.

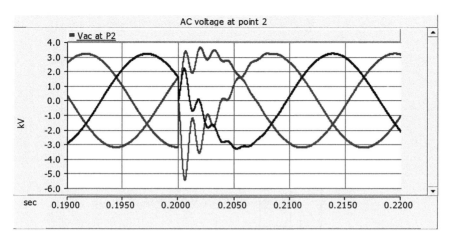

**FIGURE 2.7**
AC voltages measured at Point 2.

In order to show the FR effect, we have built the following simple PSCAD case as shown in Figure 2.10. The breaker on Phase A opens at $t = 2$ sec. The system is 13.8 kV and the inductances are intentionally used in a way so $X_L = X_C$. Figure 2.11 shows the voltages measured at each phase. Right after the single pole operation of the breaker the healthy phases will experience a maximum around 15 kV while the normal maximum voltage in this system is $(13.8/\sqrt{3}) \times \sqrt{2} = 11.26$ kV.

Figure 2.12 shows a distribution feeder that is experiencing unbalanced (single phase and double phase to ground) faults. The step-down transformer

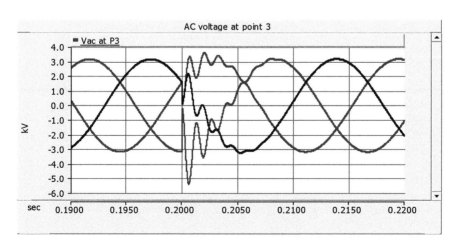

**FIGURE 2.8**
AC voltages measured at Point 3.

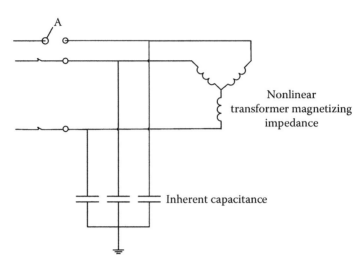

**FIGURE 2.9**
Ferro resonance.

is rated at 12.4/4.0 kV, 100 kVA ΔY. Figures 2.13 and 2.14 show the effect of single phase to ground and line-to-line to ground fault on system voltages. In the aforementioned figures the voltages of Phase A and Phase B for line to ground conditions will be swollen and also the same for Phase C in case having the fault on Phases A or B. Now we try to reconfigure the circuit and add a ground to the neutral point at the secondary of the transformer. Comparing the results in Figures 2.13 and 2.15 also Figures 2.14 and 2.16, it can be seen that the rise in healthy phase voltages is not as severe as the case without grounding.

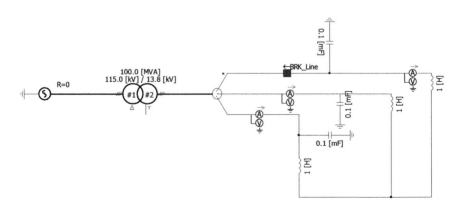

**FIGURE 2.10**
FR circuit.

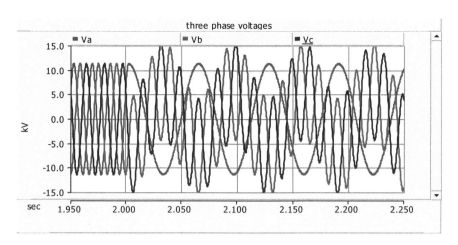

**FIGURE 2.11**
Three-phase voltages, line-to-line or unbalanced faults.

### 2.1.4 Current Chopping

Consider the following case built in PSCAD illustrated in Figure 2.17. The objective is to show when the system breaker is opened, there will an undamped energy circulating inside the filter circuit. The filter is built in a way that $(1/\omega C) \gg \omega L$, therefore the normal current inside the filter is very small. When the switch is opened, the inductor continues its current and the following relationships apply.

$$\text{Conserved Energy} \Rightarrow \frac{1}{2}Ce^2 = \frac{1}{2}LI^2 \Rightarrow e_{transient} = I\sqrt{\frac{L}{C}}$$

$$e_{normal} = L\omega I \Rightarrow \frac{e_{transient}}{e_{normal}} = \frac{I\sqrt{L/C}}{I_{(1)}I} = \frac{\sqrt{L/C}}{I_{(1)}}$$

The voltage is proportional to the magnitude of current chopped and to the surge impedance of the circuit being switched.

$$e = \sqrt{\frac{L}{C}}i = \sqrt{\frac{\omega L}{\omega C}}i = \sqrt{X_L \times X_C}\,i = X_L\sqrt{\frac{X_L}{X_C}}i$$

$$\sqrt{\frac{X_e}{X_L}} = \frac{\omega_n}{\omega} \quad \omega_n = \frac{1}{\sqrt{LC}}$$

$$\sqrt{\frac{1}{\omega^2 LC}} = \frac{1}{\omega\sqrt{LC}} = \frac{\omega_n}{\omega}$$

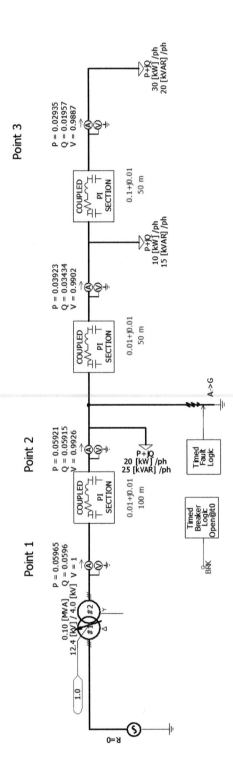

**FIGURE 2.12**
Distribution circuit to analyze unbalanced faults.

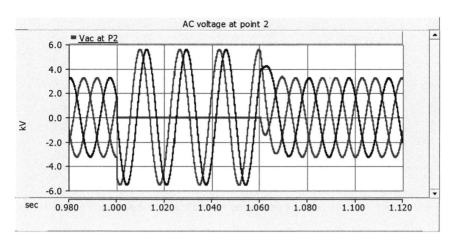

**FIGURE 2.13**
Single phase to ground fault ungrounded transformer.

$$e = \frac{\omega_n}{\omega} X_L i$$

Transient voltage $e$ is larger than the normal voltage by $\omega_n/\omega$ (Figures 2.18 and 2.19).

Other conditions may cause overvoltages, such as contact with higher voltage conductor and cogeneration. Recently with rises in solar integrations, in case of having minimal loads and a very sunny day, the feeder prevalent of solar generation may experience dangerous voltage levels.

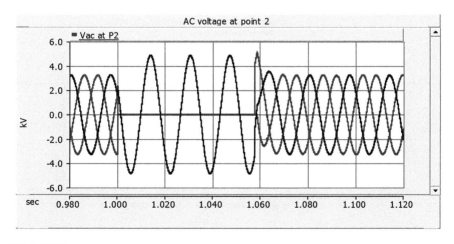

**FIGURE 2.14**
Double phase to ground fault ungrounded transformer.

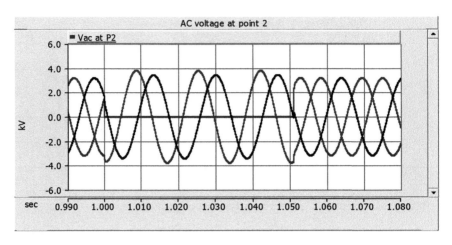

**FIGURE 2.15**
Single phase to ground fault grounded transformer.

### 2.1.5  Solution to TOVs and Arrester Devices

In order to protect equipment from damages caused by TOVs, surge arresters are placed to alleviate the large voltages caused by fast transients such as lightning. The arresters are nonlinear resistances with volt-amp characteristics as shown in Figure 2.20. The figure is showing the characteristics of two different types of arresters that are commonly used in industry. The top graph is for a metal oxide varistor (MOV), and the bottom one is for an older version, which is silicon carbide. Those devices both have nonlinear characteristics. In normal operating conditions in the case of silicon carbide,

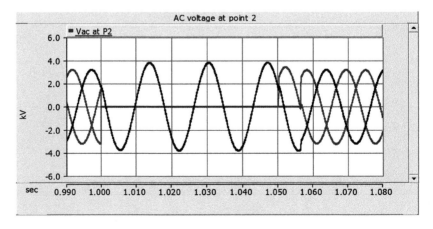

**FIGURE 2.16**
Double phase to ground fault grounded transformer.

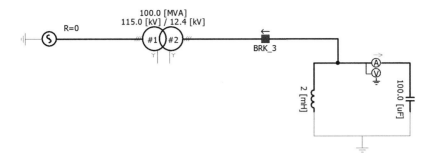

**FIGURE 2.17**
Filter circuit for current chopping effect.

the gap existing inside the device will conduct the surge to ground and acts as a very small resistance. The MOV device performs exactly the same way, but the nonlinearity of the device in this case is a lot higher compared to silicon carbide. Therefore, there will be no current passing through the device in normal conditions. The MOV device does not spark over to conduct power frequency currents. The MOV device will transition in and out of the transient very smoothly. Therefore, MOV devices have the following advantages

- Improved duty cycle performance
- Doubled energy-withstand capability
- Improved protective characteristics
- Superior overvoltage reseal capability
- Greater reliability

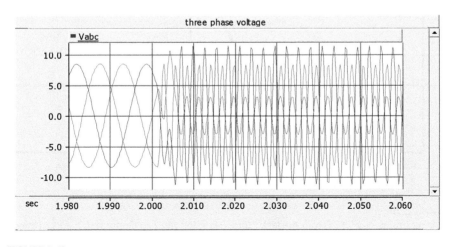

**FIGURE 2.18**
Risen voltages due to switch off filter.

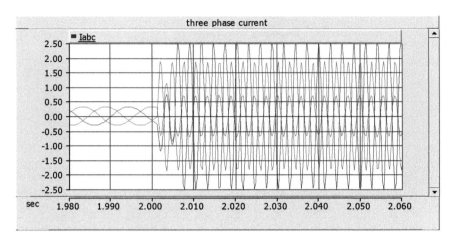

**FIGURE 2.19**
Risen current circulating inside the switched off filter.

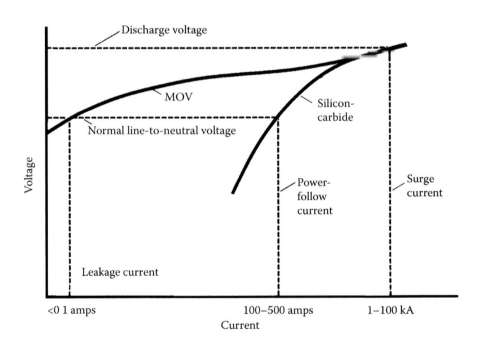

**FIGURE 2.20**
Silicon Carbide vs MOV. (From Boyle, J. "Understanding Zig-Zag Grounding Banks." *Georgia Tech*, https://www.l-3.com/wp-content/uploads/2014/04/Georgia-Tech-Zig-Zag-Grounding-Transformers.pdf)

In order to size arresters for any system, it is required to find the maximum continuous overvoltage (MCOV) that the arresters have to tolerate without operating. For instance, in a system with rated voltage of 34.5 kV, the MCOV will be calculated as following:

$$MCOV = \frac{1.05 \times 34.5}{\sqrt{3}} = 20.914 \text{ kV}$$

Therefore, the arresters have to tolerate around 21 kV without operating. Standard ANSI C62.22 gives a clear guidance as how to choose an arrester. The protective characteristics of the metal oxide arrester are a function of surge current magnitude and the time to crest of the discharge voltage. In order to protect equipment, the industry has defined several margins to be calculated relating to the protecting device.

As shown in Figure 2.21, standard industry practice has evolved into calculating margins of protection at three different volt-time points, corresponding to chopped wave, full wave, and switching surge withstand levels of the protected substation equipment. The protective margin is determined by comparing the equipment withstand level with the arrester protective level.

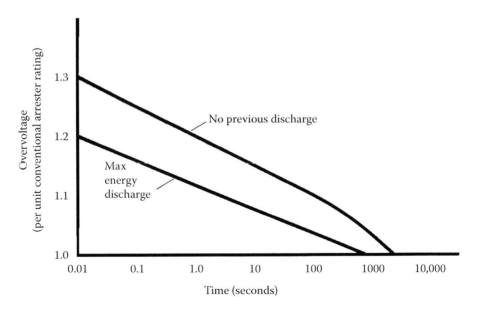

**FIGURE 2.21**
MOV capability curve. (From Boyle, J. "Understanding Zig-Zag Grounding Banks." *Georgia Tech*, https://www.l-3.com/wp-content/uploads/2014/04/Georgia-Tech-Zig-Zag-Grounding-Transformers.pdf)

$$\text{Margin}(\%) = \frac{\text{EW} - \text{APL}}{\text{APL}}$$

EW: equipment withstand

APL: arrester protective level

With that definition, there will be three margins, which are commonly used in industry.

$$\text{MP1} = \frac{\text{CWW} - \text{FOWR}}{\text{FOWR}} \times 100$$

MP1: front of wave margin of protection

CWW: equipment chopped wave withstand, typically 115% of the basic insulation level (BIL)

FOWR: equivalent front of wave response of the MOV arrester

$$\text{MP2} = \frac{\text{BIL} - \text{DV}}{\text{BIL}} \times 100$$

MP2: impulse margin of protection

BIL: basic insulation level of the protected equipment

DV: discharge voltage of the MOV arrester at a specified current level

$$\text{MP3} = \frac{\text{BSL} - \text{SSR}}{\text{SSR}} \times 100$$

MP3: switching surge margin of the protection

BSL: basic switching surge insulation level, typically 83% of BIL

SSR: switching surge protective level of the MOV arrester

Calculating the above-mentioned margins and making sure they are within the range identified by standard is called insulation coordination. This is the process of correlating electric equipment insulation strength with protective device characteristics so that the equipment is protected against expected overvoltages.

The ANSI guide for surge arresters (ANSI C62.2) recommends a minimum margin of 20% for MP1 and MP2, and 15% for MP3.

Table 2.1 shows the protecting device (MOV) characteristics for various MCOVs.

**EXAMPLE**

A 34.5 kV equipment is going to be protected by a MOV device. Find the size of the MOV device.

**TABLE 2.1**

Distribution Arresters Characteristics

Electrical Characteristics

| Voltage Rating (Ur) (kV-rms) | MCOV (Uc)[3] (kV-rms) | Maximum Equiv FOW[5] (kV-Crest) | Maximum Switch Surge[6] (kV-Crest) | Maximum Discharge Voltage (kV-Crest) Using an 8/20 μs Current Impulse | | | | | |
|---|---|---|---|---|---|---|---|---|---|
| | | | | 1.5 kA | 2.5 kA | 3.0 kA | 5.0 kA | 10 kA | 20 kA |
| 3 | 2.55 | 10.4 | 7.8 | 805 | 8.8 | 8.9 | 9.3 | 9.9 | 10.9 |
| 6 | 5.10 | 20.7 | 15.5 | 16.9 | 17.5 | 17.7 | 18.6 | 19.8 | 21.8 |
| 9 | 7.65 | 31.0 | 23.3 | 25.4 | 26.2 | 26.6 | 27.9 | 29.7 | 32.7 |
| 10 | 8.40 | 34.5 | 25.9 | 28.2 | 29.1 | 29.5 | 31.0 | 33.0 | 36.3 |
| 12 | 10.20 | 41.3 | 311.0 | 33.8 | 34.9 | 35.4 | 37.2 | 39.6 | 43.5 |
| 15 | 12.70 | 51.7 | 38.8 | 42.2 | 43.6 | 44.2 | 46.5 | 49.5 | 54.4 |
| 18. | 15.30 | 62.0 | 46.5 | 50.7 | 52.3 | 53.1 | 55.8 | 59.4 | 65.3 |
| 21 | 17.00 | 72.3 | 54.3 | 59.1 | 61.0 | 61.9 | 65.1 | 69.3 | 76.2 |
| 24 | 19.50 | 82.6 | 62.1 | 67.6 | 69.7 | 70.7 | 74.4 | 79.2 | 87.0 |
| 27 | 22.00 | 92.9 | 69.9 | 76.0 | 78.4 | 79.6 | 83.7 | 89.1 | 98.9 |
| 30 | 24.40 | 103.3 | 77.6 | 84.4 | 87.1 | 88.4 | 93.0 | 99.0 | 108.8 |
| 36 | 29.00 | 124.0 | 93.1 | 101.3 | 104.5 | 106.1 | 111.5 | 118.8 | 130.5 |

*Source:* Agrawal, K. C. 2007. "Surge arresters: Applications and selection." In *Electrical Power Engineering: Reference & Applications Handbook*, 681–719. Taylor & Francis Group: CRC Press.

Using Table 2.2 for a 34.5 kV system the BIL is 75 kV. In this condition the arrester rating will be

$$MCOV = \frac{1.05 \times 34.5}{\sqrt{3}} = 20.914 \text{ kV}$$

Calculating the standard specified margins if we use the MCOV with the size 27 kV

$$MP1 = \frac{92.9 - 86.25}{85.25} \times 100 = 7.8\%$$

$$MP2 = \frac{89.1 - 75}{75} \times 100 = 18.8\%$$

$$MP3 = \frac{69.9 - 62.25}{62.25} \times 100 = 12.28\%$$

As can be seen, this size arrester does not provide either the 20% margin for MP1 and MP2 or the 15% for MP3. Therefore, we increase the size one step and choose the MOV with a rating of 30 kV. The calculated margins will be

$$MP1 = \frac{103.3 - 86.25}{85.25} \times 100 = 20\%$$

$$MP2 = \frac{99 - 75}{75} \times 100 = 32\%$$

$$MP3 = \frac{77.6 - 62.25}{62.25} \times 100 = 24.65\%$$

When the circuit breaker is interrupting a fault, it results in arcing. The arcing happens in the interrupting medium of the breaker. During the process of interruption, the arcing medium is trying to regain its insulation property. For the interruption to be successful, the interrupting medium should withstand the fast rising recovery voltage.

### EXAMPLE WITH PSCAD

Consider the system in Figure 2.22, we are going to design and install a MOV device and compare the system results with the case without the

**TABLE 2.2**

Insulation Coordination Boundaries

| Equipment Device Withstand | Protection MOV Withstand (kV) |
|---|---|
| CWW = 1.15 × 75 = 86.25 kV (115% BIL) | FOWR (Front of Wave Response) = 92.9 |
| BIL = 75 | DV (Discharge Voltage for 10 kA) = 89.1 |
| BSL = 0.83 × 75 = 62.25 (83% BIL) | SSR (Switching Surge protection level) = 69.9 |

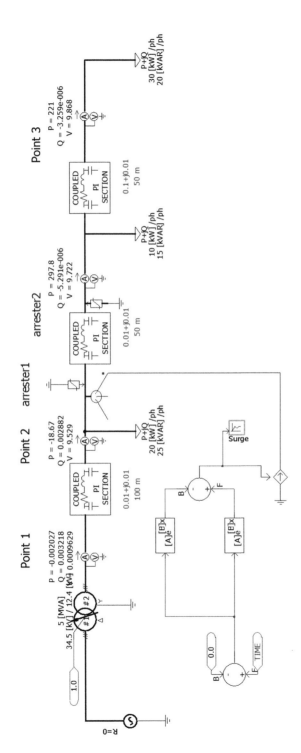

**FIGURE 2.22**
Distribution feeder protected by MOV devices.

MOV device. The system rating is at 12.4 kV, therefore the MCOV of the protecting device will be,

$$MCOV = \frac{1.05 \times 12.4}{\sqrt{3}} = 7.517 \text{ kV}$$

Comparing Figure 2.23 with Figure 2.24 illustrates the fact that addition of two 9 kV arresters alleviated ultra-high voltage impulses due to lightning in our system. Calculate the margins explained above for this 12.4 kV knowing the system BIL is 95 kV. Does the arrester size of 9 kV cover the marginal requirements?

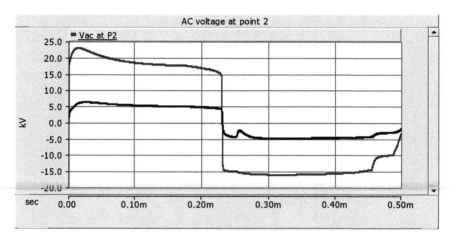

**FIGURE 2.23**
Three-phase voltages measured at Point 2.

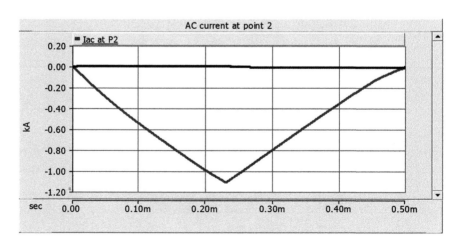

**FIGURE 2.24**
Three-phase currents measured at Point 2.

## PROBLEMS

1. What are the main transients discussed in the book? Why are they concerning?

2. What is FR and what are the possible solutions for it?

3. What is the main equipment to control TOVs? How does it function?

4. Back to Problem 3, what are the different types of the mentioned device?

5. How is it possible to make sure whether a design on an MOV device is suitable for the system? What are the inputs needed to solve a problem like this?

6. What is BIL? How is this parameter used in the system?

# 3

# Distribution Systems and the Concept of Power Quality

## 3.1 Power Quality

Power quality is a very relative concept. Something that is considered good quality for a utility may be different from what a customer expects.

In this chapter, we are going to study the following concepts related to power quality and learn how to measure quantities that will assure that a system is working with good quality.

It is expected from the electricity customers to utilize power with constant magnitude and constant frequency. But that is not always the case. There are many forms of nonlinear loads in the industry that distort the 60 Hz frequency and demand frequencies other than the main component.

Harmonics are "a sinusoidal component of a periodic wave or quantity having a frequency that is an integral multiple of the fundamental frequency."

The presence of harmonics does not mean the system cannot function. Depending on how robust the system is and how vulnerable equipment in that system are, the responses to harmonics may differ. Some of the effects that are not favorable and caused by harmonics can be

- Excessive neutral current, resulting in overheated neutrals. The odd tripled harmonics in three-phase wye circuits are actually additive in the neutral.
- Reduced true power factor (PF), where PF = Watts/VA.
- Overheated transformers.
- Nuisance operation of protective devices.
- Blown fuses on PF correction caps, due to high voltage and currents from resonance with line impedance.
- Misoperation or failure of electronic equipment.

## 3.2 Harmonics

Here to expand the idea of harmonics, we use the following example. Figure 3.1 illustrates a case of harmonic generation by power electronic devices. Without getting into the details of power electronics devices and how they function, we use the circuit built in 65 and measure required quantities related to harmonic studies. One quantity that is specified by standard is total harmonic distortion (THD).

We first need to know how to calculate THD. The following formulas are needed for that matter.

$$THD_V = \frac{V_{dis}}{V_1} = \frac{\sqrt{\sum_{h=2}^{\infty} V_h^2}}{V_1} \; ; \quad THD_I = \frac{I_{dis}}{I_1} = \frac{\sqrt{\sum_{h=2}^{\infty} I_h^2}}{I_1} \qquad (3.1)$$

Later on in this chapter, we use the related IEEE standard and evaluate if the system is functioning in compliance with the standard. Figure 3.2 makes a comparison between the harmonic rich waveform generated by blocks in Figure 3.1 and a pure wave form of $1.5\cos(\omega t)$

In this system, the source is providing for a harmonic rich load. The source current is shown in Figures 3.3 and 3.4.

Figure 3.5 shows the amount of THD of the source current and also the harmonic spectrum of the wave.

The graph showing the harmonic spectrum is actually using the Fourier transformation to change a function of $f(t)$ to its magnitude and phase as specified in following equation. The spectrum diagram is actually a phase representation of the signal.

$$f(t) = F_0 + \sum_{h=1}^{7} \sqrt{2} F_h \cos(h\omega t + \phi_h) \qquad (3.2)$$

It is also possible to have the same spectrum for angles.

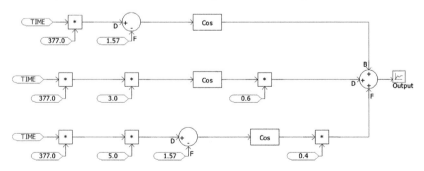

**FIGURE 3.1**
PSCAD modules for form a distorted waveform.

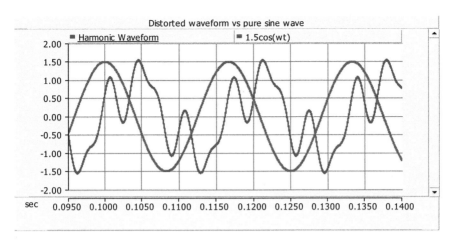

**FIGURE 3.2**
Distorted waveform versus pure wave.

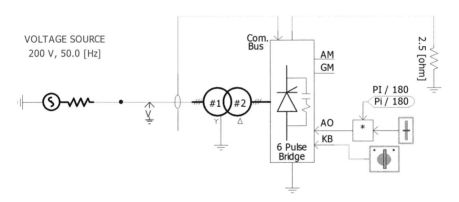

**FIGURE 3.3**
Harmonic load.

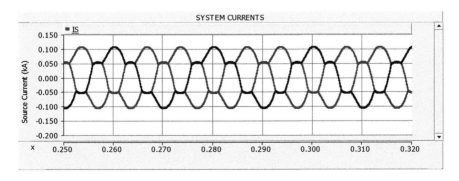

**FIGURE 3.4**
Harmonic current.

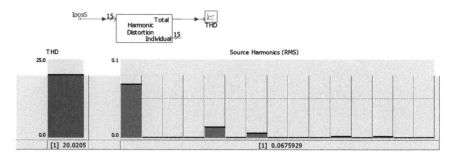

**FIGURE 3.5**
THD and harmonic spectrum of current.

### Practice

Open the PSCAD file specified here and plot the spectrum for angle. Try to explain how the module presented in Figure 3.6 works.

Here we try to do another example to analytically calculate values for THD. In a system the input voltage is

$$v_s(t) = 120\sin \omega t \, V$$

$$i_s = 85\,\sin(\omega t - 85°) + 30\,\sin(3\omega t - 30°) + 25\sin(5\omega t - 15°) + 5\sin(11\omega t - 5°)\,A$$

Find the

    a.  Fundamental root mean square (RMS) current

$$\frac{85}{\sqrt{2}} = 60.104\,A$$

    b.  Total RMS current

$$I_{RMS} = \sqrt{\left(\frac{85}{\sqrt{2}}\right)^2 + \left(\frac{30}{\sqrt{2}}\right)^2 + \left(\frac{25}{\sqrt{2}}\right)^2 + \left(\frac{5}{\sqrt{2}}\right)^2} = 66.23\,A$$

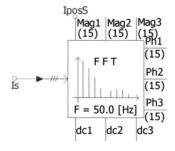

**FIGURE 3.6**
FFT module.

c. Current THD

$$\text{THD}_I = \frac{\sqrt{\left(\frac{30}{\sqrt{2}}\right)^2 + \left(\frac{25}{\sqrt{2}}\right)^2 + \left(\frac{5}{\sqrt{2}}\right)^2}}{\frac{85}{\sqrt{2}}} = \frac{27.83}{60.10} = 46.317\%$$

In general, harmonics present on a distribution system can have the following detrimental effects: (1) Overheating of transformers and rotating equipment. (2) Increased hysteresis losses. (3) Decreased kVA capacity. (4) Overloading of neutral. (5) Unacceptable neutral-to-ground voltages. (6) Distorted voltage and current waveforms. (7) Failed capacitor banks. (8) Breakers and fuses tripping. (9) Double or ever triple-sized neutrals to defy the negative effects of triplen harmonics.

This standard recommends goals for the design of electrical systems that include both linear and nonlinear loads. This procedure highlights the level of total distortion at the point that source, the linear load and nonlinear load are connected. This point is called the point of common coupling (PCC). As mentioned before, there are negative effects on the system and also loads in case of harmonics that are beyond the tolerable levels stated by standard. Therefore, this platform ensures the healthy operation of the system feeding nonlinear loads.

Table 3.1 shows the limits designed by the IEEE standard for voltages between 120 V to 69 kV.

As mentioned before, the strength of the system (in other words, the ration of short-circuit current to load current) plays an important role as to how

**TABLE 3.1**

Harmonic Current Limits by IEEE 519

**Harmonic Current Limits for Nonlinear Load at the PCC with Other Loads (for voltages 120–69,000 volts)**

**Maximum Odd Harmonic Current Distortion in Percent of Fundamental Harmonic Order**

| ISC/IL | <11 | 11 < 17 | 17 < 23 | 23 < 25 | 35 | TDD |
|---|---|---|---|---|---|---|
| <20* | 4 | 2 | 1.5 | 0.6 | 0.3 | 5 |
| 20 < 50 | 7 | 3.5 | 2.5 | 1 | 0.5 | 8 |
| 50 < 100 | 10 | 4.5 | 4 | 1.5 | 0.7 | 12 |
| 100 < 1000 | 12 | 5.5 | 5 | 2 | 1 | 15 |
| <1000 | 15 | 7 | 6 | 2.5 | 1.4 | 20 |

*Source:* Cheng, J. "IEEE Standard 519-2014." Schneider Electric. http://www.schneider-electric. com.tw/documents/Event/2016_electrical_engineering_seminar/IEEE_STD_519_ 1992vs2014.pdf

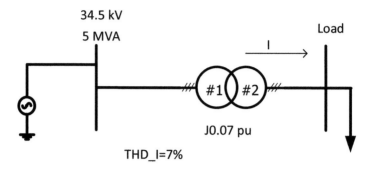

**FIGURE 3.7**
IEEE 519 compliance test.

much it can tolerate harmonics. Let us look at the following example to clarify some of the concepts that we mentioned here.

The circuit in Figure 3.7 is rated at 34.5 kV. The impedance of the line is 0.07 pu. The loads 200 kA and also there are nonlinear combinations which make the THD of current to be 7% at the PCC. Use the information in Table 3.2 to see if the system is in compliance with the IEEE 519 standard.

First, we need to calculate the short-circuit capacity (SCC) of the system. The SCC is essentially the amount of short-circuit current that the source can put to a three-phase fault at PCC. If we change the 34.5 kV to per-unit that will be 1 pu. Therefore, SCC will be calculated as follow with 0.07 pu being the impedance that connects the infinite source to PCC.

$$I_{sc} = \frac{1}{0.07} = 14.28\,pu$$

For this system $I_{base} = \dfrac{5}{\sqrt{3} \times 34.5} = 83.67\,kA$

**TABLE 3.2**

Nonlinear Load Harmonic Spectrum

| Harmonic Order | Percent of Fundamental Current |
|---|---|
| 3 | 5 |
| 5 | 7 |
| 7 | 1.6 |
| 11 | 0.33 |
| 13 | 0.22 |
| 17 | 0.11 |
| 19 | 0.09 |
| 21 | 0.02 |

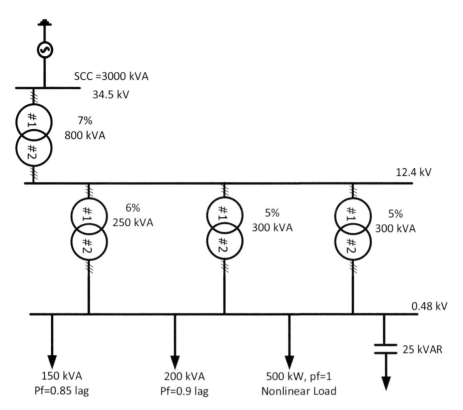

**FIGURE 3.8**
Example of harmonic assessment.

Therefore, the $I_{sc} = 14.28 \times 83.67 = 1195.342$ kA.

The ration $I_{sc}/I_L = 1195.342/200 = 5.97 < 20$

With this, we need to look at the row for <20 specified in Table 3.2. The maximum total demand distortion (TDD) cannot be more than 5%. Therefore, the system is not in compliance with the IEEE standard.

Example in the circuit presented in Figure 3.8. In this system, the nonlinear load has the harmonic spectrum provided in the following table.

- Draw the equivalent per-unit circuit.
- Find the active power passes through the high voltage transformer.
- Find the SCC at 12.4 kV.
- Check if the system is in compliance with IEEE 519.

Using the following MATLAB code and specifying the required base values, Figure 3.8 will be changed to what is shown in Figure 3.9.

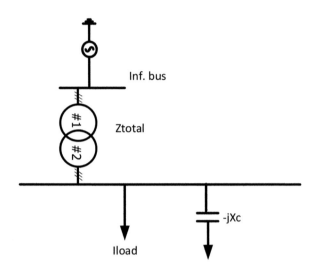

**FIGURE 3.9**
Per-unit diagram.

```
clc;
clear;

j = sqrt(-1);

%Given
Sbase = 300; %kVA
Vbase1 = 34.5; %kV
Vbase2 = 12.4; %kV
Vbase3 = 0.48; %kV

Sload1 = 150; %kVA
Sload2 = 200; %kVA
Sload3 = 500; %kW

PF1 = 0.85; %load 1 pf lag
PF2 = 0.9; %load 2 pf lag
PF3 = 1; %nonlinear load 3 pf
Scap = 25; %kVar
Ix_h = [
    3 5
    5 7
    7 1.6
    11 0.33
    13 0.22
    17 0.11
    19 0.09
    21 0.02]; %Harmonic content of Ix table [h, %]

%Find angles
```

```
theta1 = acos(PF1); % pf lag
theta2 = acos(PF2); % pf lag
theta3 = acos(PF3); % pf

% Vpu =1 => Ipu=conj(Spu)

Ifund1 = (Sload1/Sbase)*(cos(theta1)-j*sin(theta1)); %pu
Ifund2 = (Sload2/Sbase)*(cos(theta2)-j*sin(theta2)); %pu
Ifund3 = (Sload3/Sbase)*(cos(theta3)+j*sin(theta3)); %pu

%Find magnitude of all 3 currents
Ifund = Ifund1 + Ifund2 + Ifund3; %Sum of the three currents
 in pu
Ifund = abs(Ifund); %magnitude of the total current

%Find base current
Ibase = Sbase/(sqrt(3)*Vbase3); %Amps at Sbase and Vbase3

%Transformer apparent power
S_T = [800 250 300 300]; %kVA

%Transformer percent impedance
Z_T = [5 6 5 5]; %percent

%Transformer impedances in pu
T1 = Z_T(1)/100*(Vbase3^2/S_T(1))/(Vbase3^2/Sbase);
T2 = Z_T(2)/100*(Vbase3^2/S_T(2))/(Vbase3^2/Sbase);
T3 = Z_T(3)/100*(Vbase3^2/S_T(3))/(Vbase3^2/Sbase);
T4 = Z_T(4)/100*(Vbase3^2/S_T(4))/(Vbase3^2/Sbase);

%Find impedance of the infinite system in pu
Zbase = Vbase1^2/(Sbase/1000);
Zx = (Vbase1^2/(3000/1000))/Zbase; %pu

%Find impedance of the capacitor in pu
Zcap = (Vbase2^2/Scap)/(Vbase3^2/Sbase);

%Find equivalent impedances: (3 transformers in parallel plus
 the source
%impedance plus the impedance of T1)
Z1= Zx+T1+1/(1/T2+1/T3+1/T4);

%Harmonic currents in pu
Ix_pu = Ix_h(:,2)/100*abs(Ifund3);

%Total harmonic distortions:
THD_I = sqrt(sum(Ix_pu.^2))*100/Ifund

  for i=1:length(Ix_pu)
    Z1_Zcap(i)=1/(1/(Z1*Ix_h(i,1))+1/(Zcap/Ix_h(i,1)));
```

```
    Vh(i)=Z1_Zcap(i)*Ix_pu(i);
end

THD_V=sqrt(sum(Vh.^2))*100
THD_I11 = sqrt(Ix_pu(1)^2+Ix_pu(2)^2+Ix_pu(3)^2)*100/Ifund
```

In this system $I_{sc}/I_{load} = 3000/(1.2053 \times 300) = 8.29 < 20$, therefore the TDD, that is calculated as 0.48% < 5%, is acceptable by the standard. Let's look at the harmonics with the orders less than 11.

Running the above copied MATLAB code will give us the TDD of current at 480 V, to be 5.3128. The THD of voltage is 9.15.

As can be seen, the current distortion for the total harmonic is slightly higher than what the standard recommends.

The current distortion ratio for harmonics with orders less than 11, using the final line of the code, will give 5.3066 > 4. Therefore, the system is not in compliance. What would you suggest for enhancing the power quality situation on the system?

## 3.3 Voltage Quality, Sag, and Voltage Sag Solutions

Previously, we studied the effects of the frequency and amplitude of the voltage by introducing harmonics. Here, we are going to discuss another power quality issue that is always of concern. This effect is called voltage sag. When the amplitude of voltage drops to less than 0.9 pu, the quality of the voltage is deteriorated and the system is experiencing sag. The causes for voltage sag can be far away faults and load increase. For instance, in hot summer days with increase in AC loads from the houses usually distribution feeders experience low voltages. If the currents stay in the capacity of the lines, this issue can be fixed by tapping up the transformers.

The bandwidth for operating voltage at the secondary is 114–126 V. Anything less than 114 V or 0.9 pu is considered low voltage in the secondary distribution system. Motor startings in factories can cause temporary undervoltages, and the mitigation measure for that would be supporting the system with extra reactive power sources such as shunt capacitor banks or stat-var compensators.

Let's use our famous distribution feeder as in Figure 3.10 and analyze a voltage sag due to a temporary fault. At $t = 1$ sec, a three-phase fault happens at Point 3. Figure 3.11 is showing three-phase voltages measured at Point 2.

As it can be seen in Figure 3.11, for the 0.05 sec during which the three-phase fault exists, the voltages at Point 2 are experiencing a dip called the voltage sag. If we change the fault type from single phase to ground, the AC voltages

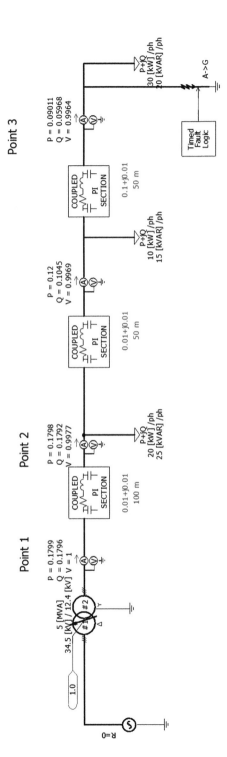

**FIGURE 3.10**
Distribution feeder for voltage sag.

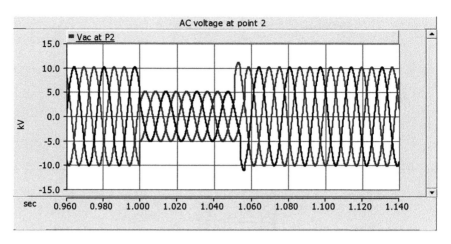

**FIGURE 3.11**
AC voltages at Point 2.

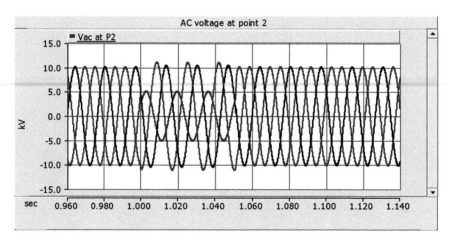

**FIGURE 3.12**
Voltage sag and swell due to unsymmetrical fault.

at Point 2 will be shown as in Figure 3.12. In this case, the phase voltage A is experiencing a sag while the other two healthy phases are experiencing a voltage swell.

Another example for voltage sag is due to induction motor starting. Figure 3.13 shows a motor starting case in a 1 MVA, 13.8 kV system. Before the machine going to steady state condition, it needs to absorb a good amount of reactive power from the grid. This will cause a dip to the voltage at the PCC. Figure 3.14 shows the effects on AC voltage and the RMS measured at the load.

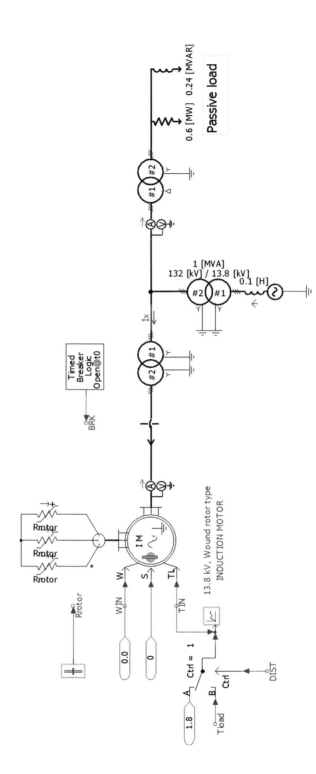

**FIGURE 3.13**
Motor starting.

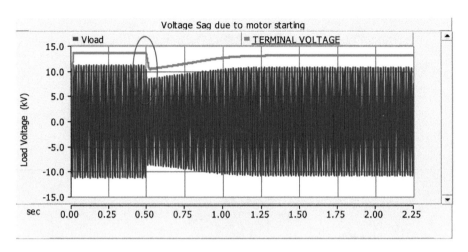

**FIGURE 3.14**
Voltage at load in a motor starting case.

### EXAMPLE 3.1

For the system in Figure 3.15. Find the voltage at PCC for Fault 1 and 2. Use a sliding fault, meaning the fault impedance is a function of length. Plot the voltage at PCC with respect to the distance that the fault appears. In order to solve this problem, we have to first ignore the loads with the assumption that the fault current is so high and the loads will be neglected. With that, the following MATLAB code helps us go through the example. As it appears in Figure 3.16, the amount of voltage sag due to Fault 2 is more than what is calculated for Fault 1.

```
clear all;
clc;

Sbase=100; %MVA
Vbase1=132; %KV
Vbase2=34.5; %KV

Zbase1=Vbase1^2/Sbase; %ohm
Zbase2=Vbase2^2/Sbase; %ohm
Ztr=0.05;
Zsc=(34.5^2/1000)/Zbase2;
Zline1=40/Zbase1; %pu
Zline2=0.8/Zbase2; %pu

x=[0:0.01:5];
Vpcc=0;
for i=1:length(x)
Vpcc(i)=(Ztr+Zline1*x(i))/(Zsc+Ztr+Zline1*x(i));
end

plot(x, Vpcc)
grid on
hold on
```

```
y=[0:0.01:5];
Vpcc=0;
for i=1:length(y)
Vpcc(i)=(Zline2*y(i))/(Zsc+Zline2*y(i));
end

plot(y, Vpcc)
grid on
hold on
```

As can be seen inside the MATLAB code that we used for this problem, we took advantage of per-unit calculations.

**Practice**

Try to change the SCC of the system and see how your curves will change.

There are several possible solutions to voltage sag, such as utilization of FR transformer: An FR transformer, or constant voltage transformer (CVT), operates in the saturation region of its B–H curve. Voltage sags of less than 30% can be compensated by them.

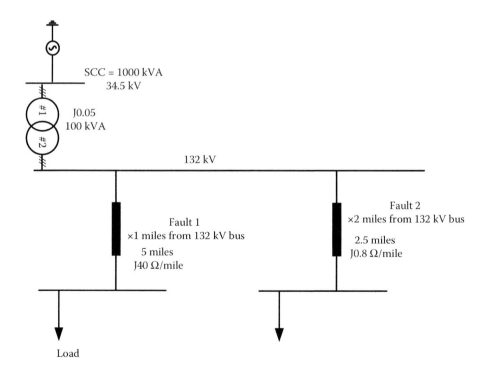

**FIGURE 3.15**
Example for voltage sag calculation.

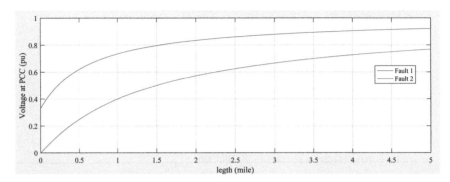

**FIGURE 3.16**
Voltage at PCC, in per-unit.

*Uninterruptible power supply (UPS)*: A UPS mitigates voltage sags by supplying the load using stored energy. Upon detection of a voltage sag, the load is transferred from the mains supply to the UPS. Obviously, the capacity of load that can be supplied is directly proportional to the amount of energy storage available.

*Flywheel and motor-generator (MG)*: Flywheel systems use the energy stored in the inertia of a rotating flywheel to mitigate voltage sags. In the most basic system, a flywheel is coupled in series with a motor and a generator, which in turn are connected in series with the load. The flywheel is accelerated to a very high speed and when a voltage sag occurs, the rotational energy of the decelerating flywheel is utilized to supply the load.

*Static var compensator (SVC)*: An SVC is a shunt-connected, power electronics-based device which works by injecting reactive current into the load, thereby supporting the voltage and mitigating the voltage sag. SVCs may or may not include energy storage, with those systems which include storage being capable of mitigating deeper and longer voltage sags.

## 3.4 Flicker

Flicker is a visual effect of the lights due to fractional harmonics in the system. The cause of the flicker is nonlinear loads, such as arc furnace or fast switchings. IEEE 141 designed a set of curves that specify borderlines of visibility and a borderline of irritation. IEEE 519 Standard: Any voltage flicker at the PCC should not exceed the limits defined by the "maximum borderline of irritation curve." If you see the lights flicker 10 times per minute, you can assume from the curve that the load in the vicinity of your house

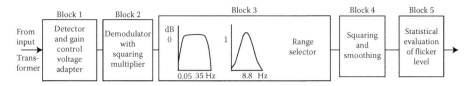

**FIGURE 3.17**
IEC flicker meter.

is having around 1.5% voltage fluctuation. Flicker intensity (FI), that is, the magnitude of the voltage variation, is determined by the power system source impedance and load peak power requirements.

Figure 3.17 shows a commonly used International Electrotechnical Commission (IEC) flicker meter. This meter is a specialized instrument that emulates a human lamp-eye-brain system. Quantifies effects of provided voltage signal on this system. The following block diagram has five main components. Block 1 emulates the section of human perception that makes moderate intensity stimuli imperceptible. It normalizes the RMS voltage of the input signal using gain-control circuit with a scaling step-response of 1 minute. It may contain two filters to help eliminate DC components and double-frequency ripple from the signal. Block 2 is designed to suppress the main frequency carrier signal (60 Hz in US). It recovers and amplifies other frequency components present in the signal input from Block 1. Frequency components corresponding to the luminosity effect of the voltage signal are among the outputs to Block 3.

Block 3 consists of weighting filters connected in series. The purpose is to strip remaining undesired components from the input coming from Block 2. High-pass filter with cutoff frequency of 0.05 Hz to remove DC component. Sixth-order Butterworth filter with corner frequency of 35 Hz to remove 100 Hz doubled carrier from Module 2 input. Band-pass filter centered at 8.8 Hz to help simulate vision system of average human observer. Specific values determined by mathematical analysis of human vision abilities and memory tendencies.

Block 4 implements the remainder of the lamp-eye-brain model. Squaring portion designed to simulate nonlinear eye and brain response characteristics. Smoothing portion is a first-order filter to emulate human brain perceptual storage effects. Output to Block 5 is called *instantaneous flicker sensation.*

Block 5 runs input from Block 4 through a sampling A/D converter and then applies statistical analysis using probability distribution and weighted sum techniques. Splits sensation into short-term and long-term severity indexes. Final output of module is objective information on flicker severity level that is independent of type and evolution of the voltage fluctuation input.

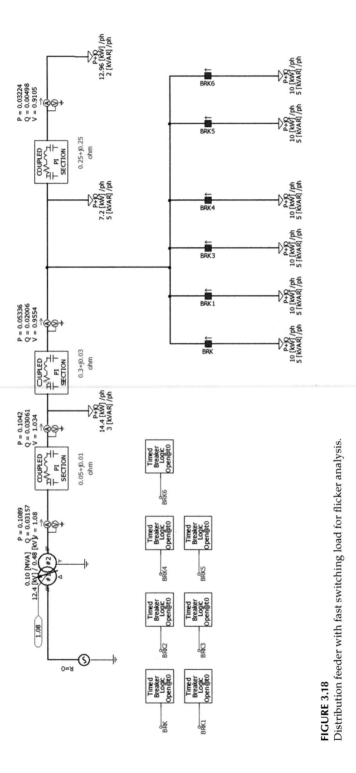

**FIGURE 3.18**
Distribution feeder with fast switching load for flicker analysis.

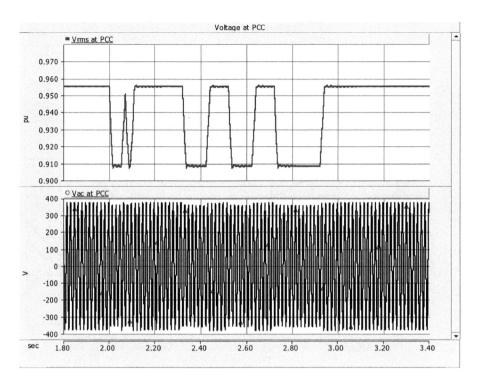

**FIGURE 3.19**
Voltage measurements at PCC.

### EXAMPLE 3.2

In order to clarify the flicker and flicker calculation, the following case is being built in PSCAD. This is a secondary of a distribution line that is sharing the feeder with a load that is being switched very fast. Figure 3.18 shows the single line diagram of the feeder along with the switching load. Figure 3.19 shows the RMS voltage and the AC voltage is measured at the PCC. The bottom figure is showing only one of the phases.

Using the bottom graph of Figure 3.19, we can calculate the FI as follows:

$$\text{FI}(\%) = \frac{\max(V_{ac\,at\,PCC}) - \min(V_{ac\,at\,PCC})}{\min(V_{ac\,at\,PCC})} = \frac{370.5 - 356.25}{356.25} = 4\%$$

The factor FI is measured for the maximum of the numerator. The event of the switching that is analyzed here is lasting for 1.1 seconds. This event is modeled on Figure 3.20 with the red dot. With 4% maximum dip and 5 dips per 1.1 seconds, this load will push the system well above the borderline of irritation.

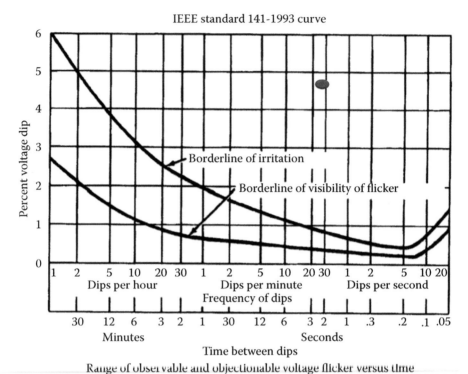

IEEE standard 141-1993 curve

Range of observable and objectionable voltage flicker versus time

**FIGURE 3.20**
IEEE flicker curve.

Practice

Design a variable capacitor to switch in and out with the load to compensate the voltage flicker.

PROBLEMS

1. Use PSCAD. For the following circuit with motor starting, use a capacitor bank that is switched at the same time with the motor and compensates for the voltage drop. What size would you choose for the cap bank? (Figure P3.1)

2. The following feeder is supplying a pure linear load and a pure nonlinear load. The impedance $R + j\omega L = 0.02 + j\omega 0.005(\Omega)$. Do a complete IEEE 519 assessment on the feeder. What is the THD of voltage at the substation? (Figure P3.2)

3. One of the ways to alleviate harmonics is by adding filters. What types of filters are common in industry? Research and provide a list.

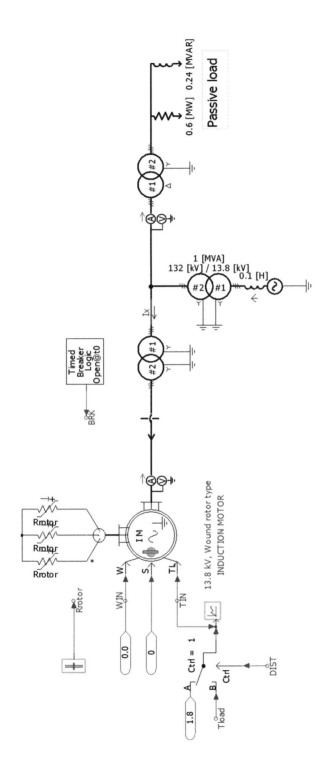

**FIGURE P3.1**
Motor starting problem.

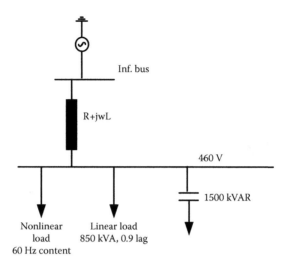

**FIGURE P3.2**
Harmonic evaluation.

Harmonic Content

| h | Nonlinear Load (A) |
|---|---|
| 5 | 700 |
| 11 | 500 |

4. Use PSCAD. For the circuit in the figure of Problem 2, design a filter that alleviates the nonlinear load harmonics effect.

# 4

## Active Distribution Systems and Related Concerns

## 4.1 Renewable System Integration and Their Impacts on Distribution System

Renewable energy resources have become an integral part of distribution systems. This proliferation of alternative energy is changing the paradigm of design and operation. Here, in this chapter we utilize some of the models provided in PSCAD for wind and solar energy and try to understand their effects on operation of distribution feeders.

The main reason for utilizing the alternative and green sources of energy is to preserve the planet and natural resources by reducing the environmentally polluting gas emissions common to coal-based power plants. Due to the variable nature of the wind and sun, renewable energy sources can be considered as generators with variable power outputs. The mechanism of power generation from renewable sources is different from traditional sources of power such as synchronous machines. Solar and wind power resources are integrated using power electronic switches (inverters). We use the concept of renewable energy integration in distribution system along with available models to show some of the impacts that renewable energy may have on distribution feeders. Proliferation of renewable sources can cause an imbalance between load and generation and, therefore, an undesirable backflow of power to the transmission system from the distribution system. Also, the variable nature of wind and solar energy can alter system parameters in unfavorable ways and cause power quality issues. The excessive power generation from wind and solar sources, if not foreseen and controlled properly, can cause operational issues on the grid such as rises in the voltage, excessive operation of voltage regulators, and protection related malfunctions.

Traditional energy management systems (EMS) mostly represent distribution feeders as a downstream, passive load. With the advent of distributed generations, however, transmission utilities may experience large reverse power flow through substation transformers and back-feeding to high voltage (HV) systems. As the penetration levels of renewable and alternative energy sources in the distribution system increase, it becomes more import to enhance the control methods applied to the distribution systems and increase the number of monitoring/representation points (e.g., in EMS).

Renewable energy sources are inherently intermittent. Similar to other natural phenomenon, the power generated by renewable energy sources will include a certain level of unpredictability despite the complex forecasting tools available today. The intermittency becomes an obstacle in controlling the output power of these sources, exposing the system to large and sudden power flow fluctuations and voltage excursions.

Wind farms and solar Photo Voltaic (PV) sources are among the commonly deployed renewable energy sources at distribution levels. Affected by the change in the wind speed and/or variations in solar radiation, these generation sources operate in an intermittent manner.

Figure 4.1 shows a 4.16 kV distribution feeder with solar integration. The solar module can generate 0.1 MW at maximum. Here, we are analyzing the cloud effect on the solar generator with solar radiation in Figure 4.2. Figure 4.3 shows the RMS voltage at the PCC. Figure 4.4 shows the active power measured at the source. As can be seen during the highest periods of power generation, the backflow in the feeder can reach around 300 kW. Maximum RMS voltage on the feeder can be around 2%. Conventionally, distribution feeders are built to operate in radial condition, and the backflow power can cause malfunctioning on the feeders' protection.

The RMS figure of the voltage is showing that the quantity has 11 dips in 9 seconds. Therefore, it can be assumed that there will be almost 1.22 dips per second. Looking at the largest dip, it will follow this calculation:

$$\frac{1.232 - 1.018}{1.018} = 0.51\%$$

We bring the flicker curve here again; the orange dot is showing the operational place at the 4 kV bus, and it is above the level of visibility.

A distribution system by nature is rich in harmonic contents. Therefore, the high THDs, as shown in Figure 4.5, may not be a big concern. With the behaviors shown above, it is required to compensate the intermittency with storage devices. The storage device, again, is a power electronic-based piece of equipment that can be utilized for voltage regulation, intermittency alleviations, and power quality enhancement. The amount of flicker that exists in this experiment is illustrated on Figure 4.6.

Strategic investments in enabling technologies can greatly increase the distribution networks ability to utilize an increase in distributed genration (DG) integration. There are research studies and ideas on how to increase the capacity of the renewable generation on a distribution

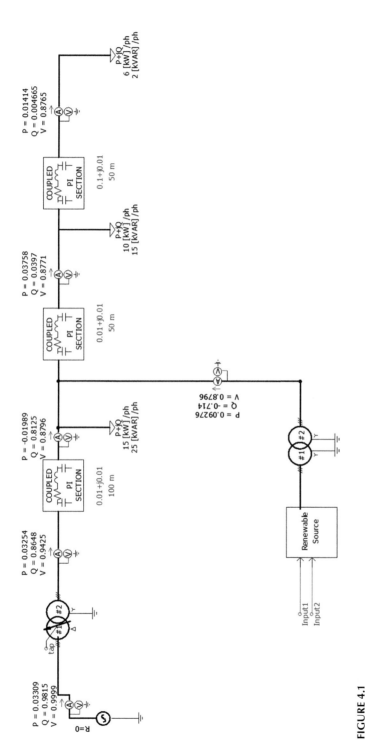

**FIGURE 4.1**

4 kV distribution feeder with solar integration. (Adapted from Muljadi, Ed, Singh, M., and Gevorgian, V. "PSCAD Modules Representing PV Generator." *National Renewable Energy Laboratory;* Manitoba HVDC Research Centre Inc. November 12, 2016, *Photovoltaic Example.* Manitoba Hydro International Ltd.; and Manitoba HVDC Research Centre Inc. November 12, 2016, *Type 2 Wind Turbine Generators.* Manitoba Hydro International Ltd.)

**FIGURE 4.2**
Solar profile 1.

feeder while maintaining the quality of service. There are simplistic and revolutionary methods to achieve the mentioned objective. The following subsection presents some of the benefits and/or shortcomings of the proposed methods to increase the DG and maintain the quality of power delivery.

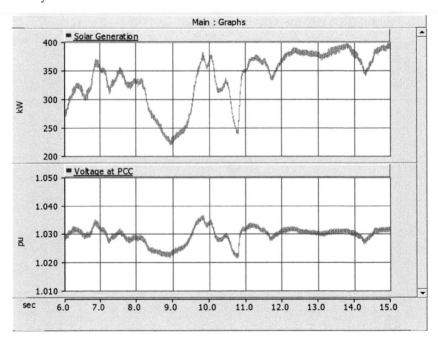

**FIGURE 4.3**
Voltage and active power at PCC.

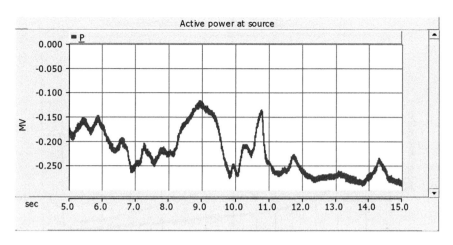

**FIGURE 4.4**
Active power at source.

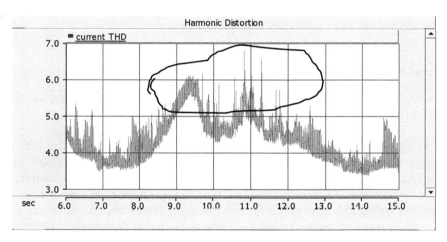

**FIGURE 4.5**
THD of the current at 4 kV.

## 4.2 Capacity Increase by Upgrading Conductors

Increasing the capacity of the lines inside distribution feeders can help alleviate the issues that might come up with high penetration of renewable energy. This is done by reconductoring the lines to improve the feeder power transfer capability. Each renewable source can be installed in a properly sized radial feeder that is served from the local HV or medium voltage (MV) substation. Although this is a practical solution, it is inherently expensive and limits the

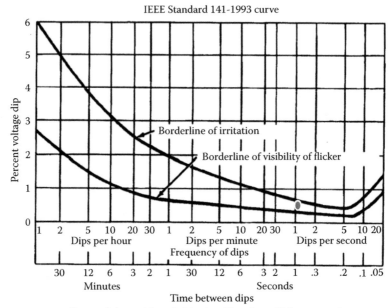

**FIGURE 4.6**
Flicker analysis. (From IEEE. 2005. "IEEE Recommended Practice for Measurement and Limits of Voltage Fluctuations and Associated Light Flicker on AC Power Systems," IEEE Standard 1453-2004.)

installation of renewable resources, such as wind and solar, at the end of the feeders. In case of weak feeders with low loads, the conductors between the renewable source and the HV/MV substation should be upgraded to increase the capacity of the feeder and, therefore, provide more opportunities for solar or wind installation. The cost of reconductoring may surpass the benefits of renewable integration and capacity increase. Therefore, this cannot be identified as the optimal solution.

## 4.3 Optimal Placement Planning

Proper placement of an individual renewable source along a distribution feeder may solve and mitigate several challenges associated with its operation. However, this may not be a good solution when the number of already installed resources are high in a feeder. Optimal power flow (OPF) can be utilized to assess the best placement for installation of resources. This is, however, a single point analysis and would have to be repeated each time a DG unit is connected to the system.

## 4.4 Voltage Control

One of the major concerns in renewable energy installation is the impact of new generation on the system voltage. Voltage regulators are responsible for enhancing the feeder voltage profile. Several types of voltage regulating devices—such as tap changing transformers, reactors, capacitors, SVCs, static synchronous compensators (STATCOMS), and other devices—are used for this matter. With the addition of FACTS (flexible AC transmission system) devices, enhanced control of the grid is possible.

As an example, local voltage rise can be mitigated with the addition of a reactor at the PCC and capacitors at the substation. Although this is an inexpensive solution for increasing capacity, it will cause an increase in the active power losses of the distribution grid.

## 4.5 Redesigning the Network to a Mesh System

As more renewable resources are introduced, the power becomes bidirectional, much like the mesh networks of the HV power grid. The effects of bidirectional power flow create their own problems.

Another meshed option may be balanced, looped secondary subfeeders, connecting consumers and producers. Despite the increased fault values, one of the advantages of such a system is for isolation of the faults. If a fault is on the main feeder, the normally open breakers and closed breakers can be rearranged to keep as much of the system energized as possible. If the fault is inside one of the loops, the fault is localized and other loops are not affected. If the loop is equipped with fault locating relays, the fault can be further isolated, splitting the loop into two radial subfeeders. Studies show that for a secured feeder with a maximum of 47% penetration of DG, the feeder could be increased to 67% penetration, depending on the number of secondary loops. The authors verified that the looped topology is capable of service continuity and also argued that the cost savings due to the reduced power losses make it comparable to that of the secured feeder. This point will have to be verified by future research and field implementation.

## 4.6 Applications of Battery Storage in Power System Improvement

Energy storage can be an instrumental part of an integrated distributed generation system. Depending on the size and type of the applied energy storage, there will be different types for such devices. These device applications can be considered in load leveling, peak-load shaving, and the dispatch during service interruptions. Energy storage devices can be utilized in short-term and long-term applications.

Short-term energy storage or power storage systems are designed and optimized for the delivery of high amounts of power over short periods of time, such as seconds or less. These devices find their application in improvement of transient stability, such as when a power system is subjected to a disturbance. Due to the intermittent nature of renewables, operation of power storage systems in parallel with most renewable generations can improve overall system behavior and the quality of power delivery.

A superconducting magnetic energy storage system (SMES) consists of a high conductance coil which stores energy in the magnetic field created by the flow of direct current. SMESs are capable of discharging large amounts of power in a small period of time. Due to the fast response of the SMES, it is capable of injecting or absorbing power to compensate for power fluctuations in the distributed system. This becomes useful in hybrid systems where generators cannot respond quickly to the fluctuations in the load. An example of this can be seen in fuel cell systems where generated output is not able to match the transient shocks due to sudden changes of real and reactive loads in the system.

Another type of fast storage system is flywheel technology, which has found a role in the regulation of the system's frequency due to its high efficiency and immediate response time. As frequency increases in the system, the flywheel absorbs energy and acts as a load to lower the frequency. When the frequency decreases, the charged energy can then be discharged into the system, acting as a short-term generation source in order to raise the frequency to nominal levels.

Supercapacitor energy storage systems (SESSs) are another type of storage capable of quickly releasing significant amounts of power on demand. An SESS is rarely used alone in a system since its energy density is low compared with other storage devices. SESS can be combined with other storage systems to provide high power density. For example, although battery (or chemical) energy storage systems (BESSs) have relatively high energy densities, they cannot be charged and discharged very quickly. To counterbalance the weakness of both BESS and SESS, these systems are placed in parallel position to provide both high energy density and high power density.

Energy management storage systems are implemented and optimized for the dispatch of power over longer periods of time. These periods range from 15 minutes to multiple hours. Long-term storage systems find their application in the areas of peak-load shaving, energy trading, integration of renewables, and/or islanded operation.

For large-scale systems, compressed air energy storage (CAES) and pumped hydrostorage (PHS) have become means of storage during off peak hours and dispatch during peak hours.

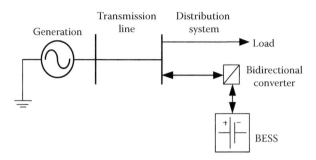

**FIGURE 4.7**
Basic configuration of BESS in a power system.

In the case of CAES, energy is stored by compression of air in a storage tank during the hours when demand is low. During peak hours when generation cannot meet the load demand, air is decompressed and fed to turbines, supporting the total generation. Efficiency of these systems is less than 70%.

PHS is essentially a hydroelectric generation unit that uses two water reservoirs, one above the other, to store energy. The water from the lower reservoir is pumped into the upper reservoir during the hours of low demand. During peak hours, the upper reservoir discharges its water through the penstock of the dam to the lower reservoir to generate power. The disadvantages of these systems are their large size and topographic and environmental limitations.

BESS is one of the most popular and frequently used means of storage due to its modularity and diverse applications. The basic configuration of a BESS is shown below. Figure 4.7 shows the implementation of battery storage at the PCC.

## PROBLEMS

1. How does renewable energy affect the voltage profile of a feeder? Explain.
2. Using the PSCAD file available for the IEEE 34 bus distribution feeder, apply the available solar module (0.2 MW) and draw the feeder voltage profile with and without the solar generation.

# References

1. 2017. "Distribution Transformer." https://en.wikipedia.org/wiki/Distribution_transformer
2. 2017. "Padmount Transformer." https://en.wikipedia.org/wiki/Padmount_transformer
3. Eaton Consultants. August 2017. "Power Distribution Systems." No. CA08104001E, www.eaton.com/consultants
4. Manitoba HVDC Research Centre Inc. 2003. *PSCAD Users Guide*. Manitoba, Canada.
5. Bokhari, A. et al. 2014. "Experimental Determination of the ZIP Coefficients for Modern Residential and Commercial Loads, and Industrial Loads," *IEEE Trans. Power Delivery* 29, 3, pp. 1372–1381.
6. 2017. "American wire gauge." https://en.wikipedia.org/wiki/American_wire_gauge
7. Peele, S. "Grounding of Electrical Systems." *Progress Energy*, https://www.progress-energy.com/assets/www/docs/business/Grounding.pdf
8. Boyle, J. "Understanding Zig-Zag Grounding Banks." *Georgia Tech*, https://www.l-3.com/wp-content/uploads/2014/04/Georgia-Tech-Zig-Zag-Grounding-Transformers.pdf
9. Agrawal, K. C. 2007. "Surge arresters: Applications and selection." In *Electrical Power Engineering: Reference & Applications Handbook*, 681–719. Taylor & Francis Group: CRC Press.
10. Cheng, J. "IEEE Standard 519-2014." *Schneider Electric*. http://www.schneider-electric.com.tw/documents/Event/2016_electrical_engineering_seminar/IEEE_STD_519_1992vs2014.pdf
11. IEEE. 2005. "IEEE Recommended Practice for Measurement and Limits of Voltage Fluctuations and Associated Light Flicker on AC Power Systems," IEEE Standard 1453-2004.
12. Gonen, T. 2014. *Electric Power Distribution Engineering*. Taylor & Francis Group: CRC Press.
13. Burk, J. J. 1994. *Power Distribution Engineering: Fundamentals and Applications*. Taylor & Francis Group: CRC Press.
14. Kersting, W. H. 2012. *Distribution System Modeling and Analysis*. Taylor & Francis Group: CRC Press.
15. AE Solar Energy. "Neutral Connection and Effective Grounding." http://solarenergy.advanced-energy.com/upload/File/White_Papers/ENG-TOV-270-01%20web.pdf
16. Medlin, G. May 2009. "Surge Protection of Distribution Equipment." *e-LEK Engineering*. http://www.ee.co.za/wp-content/uploads/legacy/01%20TD-Surge.pdf
17. Muljadi, Ed, Singh, M., and Gevorgian, V. "PSCAD Modules Representing PV Generator." *National Renewable Energy Laboratory*. http://research.iaun.ac.ir/pd/bahador.fani/pdfs/UploadFile_1989.pdf

18. IEEE Std.142-1991. IEEE Recommended Practice for Grounding of Industrial and Commercial Power Systems (ANSI).
19. Manitoba HVDC Research Centre Inc. November 12, 2016, *Photovoltaic Example*. Manitoba Hydro International Ltd. https://hvdc.ca/knowledge-base/read,article/176/grid-connected-photovoltaic-system/v:
20. Manitoba HVDC Research Centre Inc. November 12, 2016, *Type 2 Wind Turbine Generators*. Manitoba Hydro International Ltd. https://hvdc.ca/knowledge-base/topic:47/v:

# *Index*